FIRE!

FIRE!

The Story of the Fire Engine

Simon Goodenough

ORBIS · LONDON

© 1978 by Orbis Publishing Limited
First published in Great Britain by
Orbis Publishing Limited, London 1978
This edition published in 1985

Printed in Spain

ISBN: 0-85613-842-8

CONTENTS

The Enemy ...Fire

Fire engines are functional; they are also romantic. They are symbols of power, excitement and reckless courage; they represent the progressive mechanical ingenuity of man in confrontation with one of his most persistent and most feared adversaries.

For centuries men stood in line and passed buckets full of water from hand to hand to throw on the fire until their arms ached and their muscles seemed to burn with a weariness fiercer than the flames themselves. The engineering skills that had designed and worked the squirts and syringes of ancient times were forgotten for a thousand years.

Then fire fighters turned their combined energy to the hand pumps. Forty or fifty of the lustiest of them rocked to and fro, chanting at the handles, united in a common resolve to beat the fire – and to display their prowess against the boasts of rival brigades. They toiled and drank and fought each other and won the hearts of all the townsfolk. But in time they bowed reluctantly to the gleaming, polished, smoke-roaring steamers, with their teams of horses, nostrils flared, hooves sparking the streets, their pathway cleared, as they careered past, by gongs and bells and by the cries of thrilled alarm from curious citizens scattered in their wake.

Horsepower, like manpower, became history. Grinding leviathans came steam-rolling their

Left: Out of control, this fire has completely engulfed the building in a spectacular conflagration.

7

way, self-propelled, into the twentieth century, cracking the cobblestones. The new century introduced new sources of power, new materials, new horizons. Inured as its people soon became to marvels, they watched nonetheless with a suitable sense of awe as engines driven by petrol and diesel motors thundered to their vital destinations equipped with ever more elaborate gadgetry. It is easy to understand why there has been among people of all ages a fascination with engines as obsessive as our fascination with fire itself.

The Ancient Greeks believed that fire was one of the four elements that made up all matter on Earth. The assumption lasted until the Middle Ages and beyond. The Greeks had their own story of the origins of fire: they claimed that fire had been given to man as a gift by Prometheus, one of the minor deities. It was said that Prometheus took pity on man, eating his food raw and shivering in the cold, so he stole fire from Mount Olympus, which was the home of the gods. For this crime, he was punished by Zeus and chained to a lonely rock for ever.

In truth we can only assume that man's first acquaintance with fire was with lightning flashes that set the forests alight, forcing him to flee before the terrible blaze, as scared as all the other creatures. At some stage, man overcame his fear enough to turn back and learn to contain the burning embers by which he could warm himself. We may also imagine him picking up the charred body of some creature caught in the fire and tasting the meat to find that it was good to eat. The English essayist Charles Lamb once wrote a vivid and humorous description of man's discovery of the delights of roast pork and crackling when a pig was burnt to death accidentally in a domestic fire.

Groups of men living in isolated communities, migrating from place to place, constantly on the move from danger and as the seasons dictated, bore the precious nucleus of their fire in the form of glowing embers protected by a covering of clay. They rekindled the eternal flame at each resting place on their wanderings until, eventually, they discovered how to start a new fire from the friction of sticks or sparks from flint. We know that man made use of fire at least half a million years ago – remains of domestic fires dating that far back have been found with collections of animal bones strewn around the fireplace. Later, as many primitive people turned to farming and began to build settled communities, they used fire to clear the ground for crops and then for firing pottery and for working and smelting metals such as copper and bronze.

Fire also became a weapon. A fire at night kept wild animals away and protected the family. Firebrands could be brandished to frighten animals into traps; they could also be used to threaten human enemies. Settlements burned easily. The work of months might be destroyed when crops were put to the torch and buildings razed to the ground. Ships at sea were wholly vulnerable to the almost magical properties of Greek fire, which seemed, to the enemies of Greece, to scorn the extinguishing power of water.

So men learned to use fire for their own ends, yet they remained in awe of the flames. They worshipped fire and feared its strength. Fire became a focus for magic. The flames of hell terrified medieval man, yet the glowing hearth became a symbol of safety and welcome. It was, literally, a 'godsend', as Prometheus had intended, but it was simultaneously the image of damnation.

In their efforts to bring fire under control, men set element against element. They learnt that water would extinguish fire and they learnt to kick earth over the fire to exclude the air, which it needed for burning. These basic methods were not always effective, but it was a long time before men began to learn in more detail about the nature of fire and use their knowledge to develop new methods of fighting fire. Today, with more complex fires, with more and more highly flammable products unknown to our ancestors, it is increasingly a necessary aspect of the fire fighter's duty to know what he is up against. He must not only know how to fight a fire – he must understand how a fire has begun, how it will develop and what side effects it is likely to produce. This understanding is a vital part of the fireman's training and is equally an essential prelude to the story of how men have endeavoured to fight fires.

This book concentrates on the engines with which fire fighters have been equipped. It provides some idea of how those engines worked, how they developed from fairly simple beginnings, what kind of power they provided and what their attraction was to people of many centuries. There are descriptions of many of the fire engines that were the most important innovations in their time, although their importance was not always recognized by their contemporaries, most of whom adhered passionately to the traditions of the past. For example, in Britain, the steam-pump fire engine was shelved for two or three decades after its initial appearance, despite its immediately apparent advantages over the hand-pump engine. The story is coloured by some of the great fires at which these engines served and by accounts of the men who operated the pumps.

There are plenty of facts and figures and descriptions of types of pump. This is the place to find out the difference between a double-acting pump and a double-cylinder pump, between a rotary pump and a centrifugal pump. Read on, if you want some comparative figures on the distances which hand pumps, steam pumps and petrol pumps were able to throw their jets of water. How many gallons per minute could a 40-man hand-pumper throw in open competition with a ten-horsepower steamer? Could

Right: Hydraulic booms lift firemen to fight a fire in Spanish Harlem, New York.

8

a 20-man hand-pumper pump faster than two ten-man hand-pumpers? How many pounds per square inch could be shot through a 2½-inch hose at how many gallons per minute by how many Guardsmen pumping their hearts out for three minutes only?

Such important considerations were the very essence of the arguments that raged in the eighteenth and nineteenth centuries over the relative merits of different engines.

Why was Farting Annie rejected despite her initial successes? What was the dramatic story of the Exterminator, the Deluge, the Torrent, the Fireking and the Mankiller? How did the Sutherland, the London Brigade or the Hurricane excel other engines? What were goosenecks, Shanghais,

bedposts and piano-styles? These were only a few of the engines that became characters in their own time and which still enthrall us through the tales of their exploits. The names of the manufacturers themselves ring romantically across the centuries: Jynkes, Hautch, Keeling and Van der Heyden from the seventeenth century; Newsham, Lote, Hunne-man and Mason from the eighteenth century; Braithwaite, Ericsson, Hodge, Latta and Shawk from the nineteenth century. There were steamers made by Amoskeag, LaFrance, Silsby, Merry-weather, Clapp and Jones, Shand Mason; there were Mack trucks, Dederick aerials, Pirsch hook and ladders, pumpers by Ahrens Fox, Dennis, and Delahaye.

The twentieth century has contributed names as strange as any from previous centuries. The hydraulically powered snorkel raises its life-saving cage above the smoke and flames, reaching up, over and down the other side of a roof to gain access to the most out-of-the-way corners, while a curtain of water sprays out to protect the occupants of the cage. The aerialscope extends its telescopic arm high up the face of a building to combat the destructive blaze. The vastly powerful Super Pumper Complex of New York provides a water flow equivalent to an inland fireboat to ward off the threat of conflagration in that crowded city. In Chicago, the big water cannon ride on Big Mo and Big John. Turntable ladders, pump escapes, water towers, chemical and foam tenders: these are the equipment with which the modern fire fighter wages war against oil fires, domestic fires, aircraft and airport fires, forest fires, office fires, factory fires, fires caused by riot and arson. Ancillary tenders equipped with additional ladders and hoses, outfitted for salvage work or first aid duty, with generators, life-saving equipment or as command posts – all these are part of the team that backs up the pumpers and main appliances: the fire engines themselves.

Left and below: It will take a long time for vegetation and wildlife to recover from this forest fire Such fires can devastate vast tracts of land and need special equipment and fire-fighting skills.

Rioters tend to regard fire fighters as establishment figures, a branch of the authority against which they are rioting. They act accordingly, frustrating firemen in their work to save life and property. The rioters are, of course, quite mistaken in their attitude. The vast majority of people throughout the ages have recognized fire fighters as brave men risking their lives in a dangerous calling. In the days of the hand-pumpers, many girls could imagine nothing more splendid than a fireman. They were tough, often flamboyant men, who flaunted their heroism among the crowd and decorated their engines with elaborate paintings and motifs, which gave birth to the notion that anything fancy was 'all decked out like a fire engine'.

To many people, there was, indeed, more romance in the fire engines than in the firemen – or perhaps it was a combination of both. Everyone turned out to shout encouragement to the men at the handles of the hand pumps. They gazed in

Above: Trapped above the fire? The frightening peril of high-rise American buildings.
Left: Firemen to the rescue. The welcome view from the roof at a South Bronx fire in New York.

wonder at the brasswork on the steamers, with their copper cylinders and the shining harness of the horses. While the objects of their idolatry swept by amidst smoke and sparks, ordinary mortals raced to follow on foot or leapt astride their bicycles and pedalled furiously ahead of the engines.

Fire is a common enemy; to fight fire is a common cause. Fire engines are the outward sign of a universal determination to challenge the most terrible odds. They represent, in the most dramatic non-military form, the marvels of practical mechanical ingenuity directed towards an heroic purpose. This unique blend of functions ensures that men will always find fire engines fascinating.

To overcome the enemy, it is necessary first to

understand him. Fire is the rapid oxidation of matter with the simultaneous generation of heat, light and flame. When matter oxidizes, it reacts with the oxygen in the air, just as iron oxidizes and will slowly rust over a period of time.

Certain factors must exist before combustion can take place: there must be a fuel, there must be a source of heat and there must be oxygen. No matter will burn until it has reached its ignition temperature, so the source of heat is important. Once these three fundamentals have been understood, it should be clear that, to put out a fire, one of these three factors must be eliminated. The fire must be deprived of its fuel, or it must be deprived of oxygen, or the temperature of the fire must be lowered below the ignition point of the fuel. This trio of circumstances is known as the fire triangle and is fundamental to the understanding of fires and fire fighting. In recent years, the concept of the fire triangle has been extended to include the chain

reaction in which a fire builds up after the initial combination of heat, fuel and air: the fourth side turns the triangle into a tetrahedron.

Water has always been of great importance in fire fighting: it is usually cheap and in most cases it is more readily available than other extinguishing agents. Primitive man must have watched with interest as the rain put out forest fires, and he learnt to douse domestic fires in the same way. Water is able to extinguish a fire because it cools the fire by absorbing the heat energy. In terms of the fire triangle, water lowers the temperature of the fire below the ignition point of the fuel. In addition, the steam created by the effect of the heat on the water helps to smother the fire: in terms of the triangle, this cuts off the oxygen. On the debit side, water runs too easily off the surface of a burning substance, although this tendency can be corrected by adding certain chemicals. It also reacts with certain combustible metals and it conducts

electricity, so water cannot be used in fires involving these elements.

To smother a fire is also a very old way of putting it out. Dirt kicked over a small fire can quickly extinguish it. A blanket can be thrown over a fire to exclude the oxygen; if the blanket is damp it will also have the effect of cooling the fire. Chemical agents are now used as a more sophisticated means of smothering fires of many different kinds; in modified forms, chemical agents have, in fact, been used since the last century. For example, carbon dioxide (CO_2) – a gas that becomes liquid under pressure – can extinguish a fire by smothering; in the process it will also have a cooling effect on the fire. Carbon dioxide does not conduct electricity

and so can be used on fires involving electrical equipment. There are also several kinds of foam extinguishing agents. These include chemical foam, which is the reaction of dry chemicals and water; mechanical foam, which is the mix of a liquid foam agent with water and air; and high expansion foam, which forms bubbles to smother the fire – it does this by passing air through a net that holds a special solution. This high expansion foam is particularly useful for the total flooding of a confined area so that the fire is completely smothered. Foam is also highly suitable for flammable liquid fires, because of its high smothering capacity.

Dry chemicals are used to smother fires by covering the area with a fine powder which is expelled from

14

the appliance by gas. Because the chemicals are dry, they do not cool the fire, so there is always the danger that a fire might rekindle after the smothering effect has passed. Dry chemicals help to break the chain reaction that is the fourth side of the fire peril. Halogenated agents containing chlorine, bromine or fluorine are also used to break the chain reaction.

The fireman's understanding of his enemy does not stop here: he must also understand the manner in which heat is transferred from one substance to another, for this, too, will effect his approach to a fire. Every young scientist at school will almost certainly learn about the transfer of heat, but he may not link this with the drama of fire fighting.

Above: Fire can strike almost anywhere. Fire-fighters are equipped to deal with every emergency.

There are three ways in which heat is transferred: by conduction, by convection and by radiation. Conducted heat is the heat that passes from one body to another because the two bodies are touching each other: heat passes from the hotter substance to the cooler substance until both achieve the same temperature. A frying pan on a gas or an electric stove is heated by conduction. In fire-risk terms, conduction occurs when heat from a fire on one side of a wall passes through the wall to a container on the other side of the wall, thus raising the temperature of the contents of the container until those contents, too, burst into flames.

Heat transferred by convection is heat that passes from one body to another through a medium such as a gas or a liquid. A convector heater heats the air in the room, which in turn passes on its heat to the occupant. A fire might, in the same way, heat the air in a room, which in turn will heat the contents of the room until they reach their ignition points and burst into flames.

Radiation has the effect of heating substances by direct exposure to infra-red rays, visible light and ultra-violet rays. The prime example of radiated heat is the heat from the sun, whose rays upon striking an object can quickly raise its temperature dramatically to ignition point. Radiated heat is what is generally felt on the cheeks and hands when standing near a fire.

Substances may reach their ignition temperature under the effect of heat, but all substances have different ignition temperatures. For example, a lighted match may start a piece of paper burning and the paper, having a fairly low ignition point, will continue to burn after the match has been removed. The same match, or indeed the paper itself, will also start a small stick burning, but if the lighted paper or match is taken away, the stick may not continue to burn unless the flame has already got a good hold. In other words, without the help of the lighted paper, the stick will not burn unless it has reached its own ignition temperature. Coal has an even higher ignition temperature, so it is necessary to build a fire in the grate with paper and wood before the coal catches fire. So we make use of the progressively higher ignition temperatures of paper, wood and coal in order to create a fire.

A great threat to the fireman are the substances that ignite spontaneously, without being exposed to an outside source: they generate heat within themselves to bring themselves to ignition point. Dampness – from water or oils – is usually a necessary constituent of the substance that achieves spontaneous ignition. The classic example is the hay rick that is not completely cured and too loosely packed to exclude the oxygen necessary for creating the fire. Hay ricks may commonly burst into flames from spontaneous combustion. One danger still more insidious is the substance that heats spontaneously and, while not achieving its own ignition temperature, simultaneously heats up a second substance, by conduction, to that second substance's ignition temperature, so that the second, not the first, substance bursts into flames. The chances are that the subsequent increase in heat may well then raise the temperature of the first substance enough to achieve its own ignition temperature! Among those substances that can achieve spontaneous heating are charcoal, from wetting and insufficient ventilation that would otherwise have dried it; cod liver oil, being an oil-soaked organic material, when poorly ventilated; linseed oil, in oil-soaked

rags and fabrics, when poorly ventilated; fishmeal, when it undergoes an excess storage temperature with too high a moisture content; and varnished fabrics, which again may be too damp and may not be sufficiently ventilated.

Much of the experience gained through this knowledge helps in fire prevention, but it can also help the fireman to stop a fire spreading beyond its point of origin. Knowledge of the sort of materials that may be burning in a warehouse, for example, and the manner in which the fire is likely to behave are important considerations when the fireman is planning his tactics. He knows, too, that once the fire has started it will probably go through certain well-defined stages. In stage one, there is plenty of oxygen in which the fire can burn freely and heat has not yet built up dangerously high. Water vapour, carbon dioxide and, sometimes, sulphur dioxide are being given off. By stage two, the room temperature will have risen markedly and the entire room will have reached ignition point. There will be considerably more smoke and gases, including carbon monoxide, which will then become an ever-present danger to the fire fighter. It is possible, at this stage, if the room is airtight, that the oxygen will be used up and the fire will subside, but this is unlikely. By stage three, however airtight the room may originally have been, the heat will undoubtedly have cracked a window or buckled a door so that oxygen will have been admitted. This will then produce a backdraft explosion that will extend the fire so that it becomes uncontrolled. The entire building is involved in the fire at stage four.

Fire, of course, is not all the fire fighter has to face: there are the ingredients of fire – heat and flame, certainly, but also smoke and gas. Smoke is responsible for the vast majority of injuries to firemen, from inadvertent inhalation. There are several dangerous gases. These include carbon monoxide, which is colourless, odourless and vapourless and therefore cannot be seen or smelled; long exposure to carbon monoxide will result in death. Carbon dioxide overstimulates the breathing of a fireman and makes him absorb toxic gases more readily. Sulphur dioxide is extremely toxic and noxious and will cause death if not quickly avoided – even slight exposure to sulphur dioxide can cause respiratory damage and possible bronchopneumonia. Hydrogen sulphide is a more poisonous gas than carbon monoxide and can, in concentrated amounts, cause paralysis. Ammonia can be fatal, but is generally so noxious that a fireman will usually make every effort to avoid it. There are also hydrogen cyanide, which is rarely produced in dangerous quantities, and acrolein, which is also produced only in low concentrations, although it can be fatal in concentrations of as much as ten

Right: The latter stages of a fire in the Bushwick section of New York keep firemen at full stretch.

parts in a million; it is given off in the combustion of petroleum products, fat, oils and grease.

Heat produces heat exhaustion, dehydration and burns. Even protective clothing can sometimes increase the danger by causing additional dehydration of the body. Moist heat is more dangerous than dry heat. Man cannot breathe temperatures above 300° Fahrenheit, even for a short time, and the same temperature in moist heat would certainly be fatal. Pain is the only defence we have against extremes of heat; we will generally avoid the source of the heat before it becomes dangerous, if at all possible.

Most of the burns at a fire are usually produced more by flames than by heat. First degree burns affect only the outer layer of skin, which is known as the epidermis. It is this layer, less than one-tenth of the thickness of tissue paper, that assists the body in retaining its moisture and protects the body from infection and bacteria. Second degree burns affect the epidermis more deeply and cause blisters and the accumulation of fluid at the spot. Third degree burns affect the second layer of skin, which is known as the dermis. These burns leave a dry, charred area and also affect the ends of the nerves. Such deep-seated burns may cause death in more

cases than slighter burns over a wider area of the body, such as those received by a flash burn. Flash burns are very unpleasant and may cause considerable disfigurement, but they are usually only first and second degree burns. Fourth degree burns are usually deceptive and therefore dangerous. They may appear to be superficial first or second degree burns, but in fact they have penetrated to the dermis and affected the nerves without this effect being obvious to the eye. It is therefore not always evident what action should be taken. Death from burns may occur if more than 60 to 65 percent of the body is burnt, or if infection has set in or loss of body fluid has been too great – in other words, if the destruction of the skin has been such that it can no longer perform its function properly. Death by burning is therefore primarily death by destruction of the skin of the body – a dramatic indication of the vital and underestimated role played by our skin. Death can also be caused by shock.

The fire fighter is also subject to injuries from cuts and breakages, to exhaustion, and to heart failure from the stress of the work. In fact, more than one-third of the injuries received by firemen may well be associated with sprains and strains, another third from cuts and falls, one tenth from burns, and the remainder from the inhalation of various toxic gases.

In America alone, almost one million fires occur each year and most of them are in family homes.

Left: An up-to-date aerial ladder helps New York City firemen overcome a blaze.
Below: Dense populations contrast with widespread homesteads to challenge America's firemen.

The old adage holds true: the most dangerous place is your own home. Defective heating and cooking equipment is primarily responsible, especially that which ignites combustibles left nearby. Smoking-related fires also feature high on the list; so do electrical and wiring fires, rubbish fires, fires from motors and appliances, and fires from open sparks. Cooking fires and open sparks – these must have featured high from time immemorial. War, too, has caused dreadful conflagrations; earthquakes have caused fires; arsonists, unhappily, have often been responsible, as we have seen in some of the major city fires caused by riots in the last couple of decades. The mischievous spread of fire by arson is permanently one of the fire fighter's worst headaches, for it is something that no planning can prevent.

Among the great fires of history, certain conflagrations stand out and their stories are constantly retold. A conflagration is generally accepted as a fire that breaks out beyond the natural barriers of streets and individual blocks. It is a fire that grows beyond normal control, a fire that is spread by the wind and seems to take on a life of its own. In ancient times there were innumerable fires that were all-consuming and totally destructive, largely because of the flammability of the materials and the insufficiency of the methods of fire fighting. In AD 64, Rome burned for eight days and two-fifths of the city was levelled. In 798, in 982 and again in 1212, London suffered fires that were just as bad, in relative terms, as the Great Fire of 1666 in which 13,000 buildings were destroyed and more than £10 million worth of damage was done – it is hard to reckon what that damage would have amounted to in today's money. In fact, only six people were killed, but the disaster had a considerable effect on the fire consciousness of the time and led to the fire insurance of the following decades and, quite shortly, to great improvements in fire-fighting engines themselves.

In the twelfth century, Nantes and Venice were nearly destroyed. Memel suffered terrible fires in the Middle Ages, for it was common practice for foes to put cities to the flame. Boston experienced five conflagrations within a century, from 1653 to 1760. Constantinople endured many fires from the eighteenth century onwards – many of these fires destroyed up to 10,000 houses at a time. In 1812, Moscow burned for five days, put to the torch by the Russians themselves in a desperate endeavour to drive Napoleon from their capital. Nine-tenths of the city was destroyed, 30,000 houses were burned down and an incalculable amount of damage was done; this was certainly one of the most terrible of self-inflicted fires.

In 1835, the Great Fire of New York caused $15 million worth of damage and destroyed 700 buildings. Seven years later, Hamburg burned for 100 hours, killing one person for every hour it burned

and destroying 4000 buildings. The Great Fire of Chicago killed three times that number and destroyed 18,000 houses in 1871, causing $190 million worth of damage: 2000 acres in the heart of the city were totally burned out. On the same night, in the nearby forests of Peshtigo, Wisconsin, raging fires destroyed 1.25 million acres and caused the deaths of at least 1152 people. The following year, Boston suffered a Great Fire which destroyed its richest quarter and nearly 800 buildings.

The twentieth century has seen its share of destruction by fire. The San Francisco earthquake of 1906 was followed by a fire that caused $400 million worth of damage and killed 700 people. Two years earlier, New York City watched the burning of the excursion steamer *General Slocum*, with the death of more than 1000 people. Fifteen hundred people were killed in a fire in Halifax, Canada, in 1917, that destroyed 75 acres. Seven-tenths of

Above: Fire sweeps through an oil refinery.

Tokyo was destroyed and most of Yokohama in fires that followed earthquakes in 1923. These fires were particularly appalling in their effect. At one moment, 45,000 people who had taken refuge in an open park were suddenly burned up by a tornado-like firestorm. Something like 200,000 people lost their lives and possibly more than a million were seriously hurt.

War has seen some of the most terrible fires. In 1945, Tokyo suffered 84,000 dead from fire. Hiroshima suffered anything between 10,000 and 80,000 dead – the figures varied so enormously because there was little idea of the numbers exposed to the fires. Worst of all was Dresden, where firestorms caused the death of anything between 35,000 – the Russian and East German figure – and the later, independently investigated figure of 135,000.

Since the Second World War, fires have continued to rage – too frequently to recount them all or even to make a reasonable selection of them. They have burned in forest areas and bush in France and in Australia. They have burned on waterfronts, as they did in Texas in 1947 with the death of 561 people when the SS *Grandcamp* blew up. They have burned in Watts and in Washington, fired by rioters. And they have burned in oil refineries, as they did in 1975 when the tanker *Afran Neptune* was pumping crude oil and naphtha into a refinery in Philadelphia. That fire was not extinguished for eight days. The possibility of fires increases with the advent of new and dangerous combustibles, whatever the protective measures that are undoubtedly being taken. Sometime in the future there will surely be another 'Great Fire' that will go down in the history books as the worst ever. Equally certainly, firemen will risk their lives to fight their old enemy.

21

The Early Days

At first, fire fighting was largely a question of manpower rather than machines: men were needed to throw over the fire the contents of water bags made from the skins of animals, or to beat at the fire with branches torn hastily from nearby trees, or to form a bucket chain to the nearest water source. The methods changed little in Classical times and throughout the Middle Ages. Yet the Romans did know the secrets of certain kinds of engines, secrets that were lost for more than a thousand years after the fall of the Roman Empire. Knowing their expertise and the skills of other civilizations of the Mediterranean and Asia Minor as we do, it is hard to imagine that there were not engines of one kind or another invented even earlier by the mechanically adept engineers of ancient cities such as Babylon.

The first outstanding engineer whom we know for certain applied himself to the problem of throwing a stream of water into a fire by means of a pump was Ctesibus, an Alexandrian of the second century BC. The pump he invented was described by the historian Vitruvius in some detail. It was made of brass and had two cylinders, with valves to let the water in and out, and pistons fitted very closely and greased in oil so that they ran as smoothly as possible. The pistons were worked by bars and levers from above, while fork-shaped pipes led from the cylinders and

Left: Jan Van der Heyden's new fire engines go into action in Amsterdam, with willing helpers (1673).

met in a basin in the middle, whence the air pressed the water upward through the pipe.

It is fairly evident from the description handed down to us by Vitruvius that the pump made by Ctesibus had an air vessel – an essential component which enabled the water to be produced in a continuous stream instead of in a pulsating jet, as with a simpler pump. The theory of the air vessel is not very complicated, but it was not rediscovered until the seventeenth and eighteenth centuries. In simple terms, an airtight chamber is fitted to the pump; this chamber contains air at atmospheric pressure. As water flows through the outlet of the pump, some of the water enters the air vessel and compresses the air within the vessel. Thus, when the stroke of the pump slackens, the air, which is now at a higher pressure than the outside atmosphere, forces the water in the vessel out until the next stroke of the pump, when the vessel is once more filled with water, which it once again ejects when the pump stroke slackens. And so the process is constantly repeated, providing a smooth flow for the fire fighter.

There were less sophisticated methods for obtaining a relatively steady flow of water but they were never quite so satisfactory, although much more common. The ordinary force pump could be provided with a double cylinder or it could be made double-acting. The double-cylinder pump had, as its name suggests, two cylinders: as one pump stroke was made in one cylinder, the other relaxed, and as the second stroke was made, the first relaxed, so that there was always one stroke being made to push out the water. In the double-acting force pump there was only one cylinder but the stroke of the pump acted both to press water down through the pipe and to lift it up, thrusting the water through outlets above and below the plunger. In contrast, a single-acting pump is one in which the water acts only on one side of the plunger. Double-cylinder force pumps were certainly known to the Romans.

A double-cylinder pump with an air vessel was described by another Alexandrian of the second century BC, called Hero. This engine had two brass force pumps connected to one discharge pipe; it had the additional advantage of a 'gooseneck' or jointed movable tube in the discharge pipe, which enabled the jet of water to be pointed in any direction. This so-called 'gooseneck' was not rediscovered until the renewed interest in fire engines in the eighteenth century was well under way. Hero refers to the engine as a *siphos*, and in many ways it was probably quite similar to that which Vitruvius credits to Ctesibus. Pieces of similar engines have been found at Bolsena in Italy and at Silchester in Britain.

There are scarcely any worthwhile records of the practical use of these engines and there were probably very few in existence. There may have been a number of the simpler, single-acting barrel pumps of the type that were occasionally to be found in Europe during the Middle Ages, particularly in Germany, Belgium and Holland. These countries were far in advance of anywhere else during that time. The historian Pliny does in fact refer to a machine which he calls a *siphos*, but it is not wholly clear from his description what sort of machine this was. We can only tell that he must have considered it more efficient than ordinary buckets and water pipes because he recommends that, if one of these is available, it should be brought to the fire as a first measure of attack before resorting to the bucket chain. If there was no *siphos* available, Pliny further recommended the use of leather bags filled with water and connected to long pipes; the bags should be compressed by manpower in order to shoot the water through the pipes up to a reasonable height. The Romans, too, had their problems with high-rise flats and apartments in their congested cities, where building land was at a premium.

Machines, even in Roman times, were, however, in short supply, and the Romans relied chiefly on manpower and bucket chains, as did everyone else when it came to the crunch. Manpower was supplied in the main by slaves, and so there was no shortage. The body of slaves employed as fire fighters was known as the *Familia Publica* and the tribunal of magistrates in charge of it was known as the *Tresviri Nocturni*. In general the Romans had a fatalistic attitude towards fire, as did most primitive, Classical and medieval people. They all regarded fire as either a gift or a curse from the gods and reckoned that there was little they could do about it or, indeed, were meant to do about it. However, the Romans did at least have quite a degree of wealth and were keen to protect their wealth – a sense of greed is a marvellous activator against the inertia of fatalism!

The slaves were stationed around the walls of the city and at the gates during the night and were supposed to give the alarm if they spotted any sign of fire. But it was only the most confident citizens who believed that the slaves would always hold the best interests of their masters at heart. Many masters were unpopular and the slaves might leave a fire to burn a little before sounding the alarm; it was easy to turn a blind eye to a neighbourhood where loyalties were strained because of harsh treatment or a hard-held grudge. There were plenty of complaints that the *Familia Publica* was slow in reaching fires and that it seemed unwilling to take any personal risks in its efforts to put out the fires, or to save life or property. It was, therefore, not perhaps surprising that one quarter of Rome was destroyed in the dreadful fire of 6 AD.

Alarmed by this catastrophe and well aware of the weakness of the system of *Familia Publica*, the Emperor Augustus instituted the corps of *Vigiles* to protect the inhabitants of the city from the perils of fire. Rome was divided into seven fire areas and each area was allocated to one cohort of a thousand *Vigiles*; each cohort was sub-divided into ten companies. So there were 7000 Vigiles to take care of the one million inhabitants of Rome, a far better ratio of fire fighters to citizens than has been achieved by any modern city. The *Vigiles* protected Rome for 500 years.

At first the *Vigiles* were recruited from freedmen, who entered into a long-service contract; later inducements included full citizenship status after service for six years. Within 100 years of the institution of the *Vigiles*, free-born men were joining because of the high status that the corps offered. They were housed in excellent barracks, or *excubitoria,* with baths and gymnasia for keeping fit, and their job carried with it considerable social rank: they were considered to be an *élite* arm of the Roman forces. Much later, Napoleon imitated this link with the armed forces with his *Sapeurs Pompiers*. A corps of *Vigiles* also existed in Britain.

The corps itself was made up of men with specific duties. Those duties tell us a great deal about the sort of equipment that they had and their methods of dealing with fires. There were three basic duties: that of the *Aquarius*, or water man; the *Uncinarius*, or hook man; and the *Siphonarius*, or pump man. The *Aquarius* organized the bucket chain to fetch water from the nearest source and pass it in buckets from hand to hand down the line to the fire and then, in reverse, to pass the empty buckets back down the line to be refilled. Usually there was a double

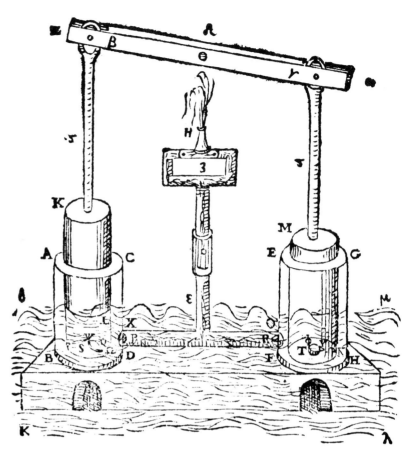

Above: The double-cylinder pump with air vessel described by Hero of Alexandria.
Far left: A fanciful sixteenth-century re-construction of Ctesibus in his study.

chain: one line to pass up the full buckets and one to pass down the empty ones. The *Siphonarius* was a bit of a specialist: he took control of the pump engine, if there was one available. We do not know how many engines the *Vigiles* had at their disposal at any one time; no one seems to have thought it worth recording. But the *Siphonarius* would no doubt have taken considerable pride in his machine and jealously guarded his authority over it as well as taking pains to keep it in good working order – he was the mechanic of the corps. The *Uncinarius* was, in fact, no less important: he wielded a long hook, with which he was responsible for dragging burning material from the roofs and walls of the houses that were on fire. Upon his strength and agility depended to a large extent the speed with which the fire could be put out.

There was other equipment, carried by other members of the corps, besides the long hook, the buckets and the pump. There were blankets which had been soaked in water so that they could be used to smother the fire. These were the *centones*. They would also be used to protect nearby buildings from the heat of the fire; by being hung over the walls of those buildings they helped to ward off

25

radiated heat. There were poles, or *perticae*, which were used with the hooks to pull down the burning house in order to stop the fire spreading to other houses. Ladders, or *scalae*, were needed for rescue work from upper stories. Wickerwork mats, or *formiones*, helped the damp blankets to smother the flames or might be laid over embers too hot to walk over so that the fire fighters might get nearer the flames. Pickaxes, or *dolebrae*, and felling axes were additional equipment for the pole man and the hook man to break into the house and disperse the fire or to create a fire break. Brooms *(secures)* and sponges *(spongiae)* were used to clear up the mess afterwards, in the manner of the modern salvage corps; although perhaps the sponges might also have been used to help to cool down the fire fighters themselves as they laboured in the heat and smoke.

The *Vigiles* performed other duties, also. Their responsibility was to prevent fire as much as to fight it, and so they had right of entry wherever they suspected a fire, a right which inevitably led to a certain amount of abuse. The Prefect also had the right to order a householder to be flogged if he felt that a fire had been caused by the direct fault or negligence of the householder. Apart from duties directly connected with fire, the *Vigiles* were responsible for recapturing runaway slaves and, a little surprisingly, for watching over the clothes of bathers at the public baths, to prevent the clothes from being stolen! There would be a fierce outcry by today's fire fighters if such duties were expected from them at the municipal swimming baths.

Far right: Fire as a weapon of war, as Jerusalem is besieged in the Middle Ages.
Below: The secret weapon! Greek fire swings the balance of naval warfare in this illustration from a Byzantine manuscript.

Of course, there was one considerable disadvantage to the *Vigiles*, in contrast to the *Familia Publica*: the *Vigiles* had to be paid out of the public pocket. This needed some persuasion, but on the whole it was seen that it was cheaper to pay out money to the corps than to lose one's entire wealth in a fire; the efficiency of the *Vigiles* was their own justification. Nonetheless, there were complaints against them, and not only because they were expensive. They had their own affiliations which often interfered with their reliability. Supported by the emperors, who were determined to keep them as their own *élite* force, they were only too happy to subject themselves to political control and reap the many rewards of loyalty to a powerful master. It has even been suggested that political manipulation of the corps may have been responsible for its failure to control Nero's notorious fire of 64 AD.

Unfortunately, so much of our understanding of the fire fighters of Classical times has to be speculation, simply because there are few references to them and those references are mostly made in passing. The letter writers and historians of the day did not often think it interesting or necessary to describe the details of an everyday event to people who knew it well enough. Their assumption of knowledge is our loss.

When the Roman *Vigiles* had ceased their duties and the air vessel of Ctesibus and the *siphos* of Hero had passed out of memory, there was a period of more than a thousand years before the reappearance of the air vessel in Europe. During this period, a strange motley of fire engines were recorded. They showed little signs of mechanical progress until towards the end of the seventeenth century; on the whole they would appear to the casual observer to be very simple or somewhat eccentric. What they do show is excellent evidence of the inventiveness

of man when his resources and understanding are strictly limited.

For the most part citizens were still armed almost solely with buckets, hooks, sticks and wet blankets. Many relied more on faith than anything else with which to combat the fire, for they continued to regard the visitation of fire as an act of God. After the Great Fire at Canterbury in 624, the monk and historian Bede related what happened: 'The great fire reached the Church of the four crowned martyrs, the casting of water would not stop it and it came rushing towards the cathedral. Archbishop Mellitus, suffering from gout, had himself carried there, and himself flaming with divine love and sanctity prayed, whereupon the wind miraculously changed and blew the fire away.'

More practical remedies laid emphasis on the prevention of fire rather than last-minute – and perhaps over-optimistic – methods of putting it out. William the Conqueror introduced the idea of a curfew in 1066 but he was not the first to do so; Alfred the Great had established a similar precaution to a limited extent several centuries before. It was, however, from William's reintroduction that we get the modern word 'curfew', from the French *couvre feu*, or 'fire cover': a metal cover was placed over the fire to exclude the air from it and so put it out. Under this law of curfew, all fires had to be extinguished at night.

Medieval accounts of fires in Britain are numerous and so are the accounts of repeated attempts to control and prevent fires. In the twelfth century, London's first Lord Mayor, Henry Fitz-Alwyn, attempted to impose strict building limitations on the city in an effort to stop the buildings becoming too crowded. In the fifteenth century, wooden chimneys were forbidden in London. The sixteenth century made few advances. Fires were still fought 'by the help of God and good, well disposid people as the mayer, the shyrevys and other good cytyzynes'. In an emergency, the town might turn to anyone for help: the brewers, for example, were always useful, with their large barrels which could serve as water carriers. There are records revealing that in some towns it was the brewers' established duty to bring their drays loaded with barrels of water to the scene of the fire. We also have an old illustration of a barrel slung on a pole being carried between two men to a burning church.

Private citizens, even then, were supposed to be able to supply their own buckets. At Winchester in

Above: Fire-hook and bucket at 'the lamentable spoile by fire upon the Towne ot Tiverton in Devon, April 3rd, 1598', from Wofull news from the West-parts of England.
Far right: Syringe, hooks and buckets on the wall, from Agricola's De Re Metallica *of 1548.*
Below: Fire squirt in action from an illustration in Cyprian Lucar's Treatise of 1590.

1574 it was the responsibility of each person to have his own leather bucket on pain of a fine of six shillings and eightpence. At Bristol in 1586 members of the Common Council were expected to have as many as six buckets each. In the event of fire, everyone turned out to help. In 1561, there were 500 people involved in bucket chains trying to save the burning steeple of St Paul's in London. But a disastrous lesson was learned in the Devonshire town of Tiverton in 1598, when 400 houses were destroyed and 50 people were killed because there were not enough buckets to go round even though it was market day and there were hundreds of people who could have helped had they had the means to do so.

As to engines, there are virtually no records before the end of the fifteenth century. One of the earliest wheeled fire engines was constructed by a goldsmith of Augsburg in Germany who was called Alton Plater. So far as we can judge, this engine was simply a large squirt set on wheels. The squirt was merely a cylinder with a nozzle into which was pushed a plunger to press out the water; the plunger was then drawn back, water was poured again into the cylinder, and the plunger was pressed home once more, while the whole contraption was aimed in the general direction of the fire. We have no idea how effective the engine was, but he might guess that it could send a jet of water quite a long way since, being set on wheels, it was clearly fairly large.

Another squirt is to be found in an illustration from the book known as *De Re Metallica* which was published by Rudolphus Agricola about 1550. The sketch shows a complete set of fire-extinguishing apparatus in the shop of a metallurgist. Besides the syringe, or squirt, which rests on hooks on the wall, there are also a sledge hammer, two 'preventers', or fire hooks, and three buckets, probably made from leather. Littered around the floor are several more instruments in the process of construction by the metallurgist, as well as a broom and what might be a wicker mat, possibly used to help smother fires in the same way as the wicker mats used by the Roman *Vigiles*. The shop of the metallurgist would be more susceptible than most to an outbreak of fire so his was a wise precaution, but perhaps other shops, too, had similar fire points.

Agricola's squirt was made of brass. The ordinary squirt of this kind was usually operated by more than one man. It had two handles, one on either side, and was held by one man on either side while a third pressed home the plunger; as he did so, the other two would brace themselves against the thrust and attempt to aim the squirt. After firing the squirt, the two sidesmen would push the nozzle down into a large basin of water which was kept constantly filled by buckets; the third man would then draw back the plunger to siphon the water up into the squirt ready for the next thrust. It was a slow and heartbreaking method of fighting a fire –

no wonder they tried to experiment with one or two less orthodox engines! Even so, such squirts were in use well into the seventeenth century.

The next engine of which we have good evidence was devised by Cyprian Lucar in 1590. His *Treatise Named Lucarsolace* was published in London that year and in it there was an illustration of a large fire squirt operated by a screw thread. Instead of the plunger being thrust down the cylinder, it was wound in on a screw thread; the water was poured in continuously from a funnel on top of the front end of the cylinder. The cylinder itself was shaped conically, rather like a modern cement mixer, with a nozzle at the smaller end. Lucar himself described the machine and was clearly extremely proud of it, although we may question its efficiency and we have no evidence of it in use: 'I will set before your eyes', wrote Lucar, 'a type of squirt which hath been

devised to cast much water upon a burning house, wishing a like squirt and plenty of water to be alwaies in readiness where fire may do harm, for this kinde of squirt may be made to hold an hoggeshed of water, or if you will a greater quantity thereof, and may so be placed on his frame, that with ease and smal strength, it shall be mounted, imbased or turned to any one side right against any fixed marke, and made to squirt out his water with great violence upon the fire that is to be quenched.'

The Germans were among the most advanced fire fighters of the time and one German machine was described by a Huguenot refugee called de Caus in his book *Forcible Movements*, published in 1615. Describing the engine, which was towed around on a sledge, de Caus said that it was 'a rare and necessary engin, by whiche you may give greate reliefe to houses that are on fire. This engin is much practiced in Germany, and it hath been seen what great and ready help it may bring; for although the fire be 40 foot high, the said engin shall there cast its water by help of four or five men lifting up and putting down a long handle, in form of a lever, where the handle of the pump is fastened. The said pump is easily understood: there are two suckers within it, one below to open when the handle is lifted up, and to shut when it is put down, and another to open to let out the water: and at the end of the said engin there is a man which holds the copper pipe, turning it to and again to the place where the fire shall be.'

The illustration shows an angular connection between the main pipe stemming from the barrel and the delivery end of the pipe, and it is possible that this might have been a flexible connection to help the fireman aim his jet more easily at the fire. This very basic form of engine was quite possibly used at the time of the Great Fire of London in 1666, being supplied with water from bucket chains that were filled from the Thames in a vain attempt to hold back the spreading flames.

An engine with wheels would have been a great deal easier to draw than a sledge, except in winter in the snow, and another wheeled engine appears in John Bate's *A Treatise on Art and Nature,* which was published in 1634. In this work, Bate described several different engines. He prided himself that the use of his wheeled machine had 'been found very commodious and profitable in cities and great townes' and he gave full descriptions of 'an engine to force water up to a high place, very useful for to quench fire amongst buildings. Let there be a brasse barrell provided,' he wrote, 'having two succurs at the bottom of it; let it also have a good large pipe going up one side of it, with a succur nigh unto the top of it, and above the succur a hollow round ball having a pipe at the top of it, to screw another upon it to direct the water

to any place. Then fit a forcer unto the barrell, with a handle fastened unto the top. At the upper end of the forcer drive a strong screw, and at the lower end a screwnut. Put them all in order, and fasten the whole to a good strong frame, that it may be steddy, and it is done. When you use it, either place it in the water or over a kennel, and drive the water up to it, and by moving the handle to and fro it will cast the water with mighty force up to any place you may direct.'

The same primitive type of pumps had been used in the New World, sometimes taken over by ship or built on the spot. There are plenty of early mentions of water buckets, hooks and ladders to combat the ever-present threat of fire to the wooden houses that the first settlers in America constructed close together for safety. It was a surer protection against Indian attack to build upwards and not outwards; a settlement that was spread over a wide area could not be guarded adequately. Boston suffered the most from fire of all the early colony towns. Massachusetts and Philadelphia both took precautions by passing laws against smoking outdoors – the danger being, presumably, that you would be foolish enough to throw the ash near a house, an act that you would sensibly avoid *inside* your own house. New Amsterdam took still firmer measures. Governor Peter Stuyvesant laid down strict laws controlling the construction of houses in the settlement; the houses in any case were built of stone, which made them less of a risk. Fines were imposed in New Amsterdam, as elsewhere, for disobeying the laws and the money received from the fines was used to buy fire-fighting equipment. Not all fires, however, were accidental. Arson was a common crime and the

Right: Citizens had little more than fire squirts with which to fight London's Great Fire of 1666.
Below: This solid-looking pump from de Caus's book, Forcible Movements *(1615), has a swivel joint.*

punishments were severe; it was not unknown for the arsonist himself to be burned alive at the stake! Volunteer fire groups were formed to keep watch in the streets at night and they gave warning of fire with rattles. In consequence they became known as the Rattle Watch.

Although there were probably fire engines in America earlier than the middle of the seventeenth century, the first ones to be recorded there were built about that time by Joseph Jynkes, an iron-worker, who was commissioned to build 'ingines to convey water in case of fire'. We do not know exactly what kind of engines these were. Perhaps they were simple syringes or squirts; perhaps they were more advanced tub-like pumps with handles.

It was also about this time that a considerable development occurred in Germany – the reappearance of the air vessel in the engine made by Hans Hautch in Nuremberg in 1655. This was the secret of Ctesibus, at last rediscovered, and yet the rediscovery remained unsung until the beginning of the next century. Once again it was possible to produce a constant stream of water with which to fight the fire, by means of the large copper air cylinder. A model of a similar engine to Hautch's, with a double pump, still exists from about the year 1680. The model might well have been used by the manufacturer to help him sell his full-size engines. In the model there are two swivel joints on the delivery pipe, which enable it to be turned in any direction on both the horizontal and the vertical planes. Engines of this kind were not always on wheels; and besides being drawn on sledges, like John Bate's engine, they might be carried on poles by any number of men, either on their shoulders or at arm's length by their sides.

Although it was these kinds of engine that were slowly being made available to come to the aid of the bucket chains in cities throughout Europe, America and Britain in the latter half of the seventeenth century, there were few enough of them and, when London suffered its worst-ever fire in 1666, the fighting was done mostly with buckets. At first many Londoners failed to realize the enormity of the peril. 'Pish!' said the Lord Mayor of London, when woken in the middle of the night to view the growing fire, 'A woman might piss it out.' And with that he went back to bed. By morning the Great Fire was a mocking inferno. It was not the first great fire that London had endured. There had been one almost as terrible in 1212, which itself had been known as the Great Fire of London until its still more terrible successor. There had also been a fire earlier in the same century, in the year 1633, in which the citizens, trying to escape from the fire which raged on one bank of the River Thames, were caught as they crossed London Bridge when the fire jumped the river and set ablaze the bank to which they fled.

Despite these warnings, the lesson had not been learned. The crowded, overhanging houses of the city were a fire trap and protection against fire was strictly limited to buckets, hooks, squirts and the rare engine. Two-thirds of the city was destroyed in four days of continuous burning. Nearly 90 churches were ruined and about 13,000 houses. While these burned, the citizens argued among themselves as to how they should tackle the fire and where they should create a firebreak, for none wanted to have their own houses pulled down in order to save the spread of the fire. Even as they argued, the fire spread and burned those houses they had refused to pull down. Eventually the king himself intervened and houses were blown up with gunpowder in order to create a path over which the fire might not leap. Boatmen made their fortune during those panic-stricken days as citizens struggled to save what they could from their houses and flee across the river. The boatmen were able to charge exorbitant rates to those who were prepared to pay almost anything to save their belongings and their own lives. Men came from far outside London to gaze and to make what profit they might snatch from the disaster.

In the end, the fire did at least profit the citizens themselves. It brought them to their senses. There arose a new awareness of the danger of fire and a new determination to prevent such a disaster from recurring. Similar fires in Boston and elsewhere in America – there was an eight-hour fire in Boston in 1711 which destroyed 100 buildings and killed 12 people – made the people of the New World equally determined to prepare themselves more adequately, but at first they were content to import most of their machines from England.

In the year after the Great Fire an opportunist physician named Dr Nicholas Barbon hit upon the idea of insuring the citizens of London against the

Above and below: Two fire insurance marks from the days when fire insurance offices maintained their own fire brigades. The Hand-in-Hand Fire and Life Insurance Society (above) with the number of the policy below the clasped hands, was established in 1696. The Union fire insurance office (below) was established in 1714.
Bottom right: A Keeling tub pump of the 1670s, used in Dunstable until the nineteenth century.

risk of fire. In return for a premium, he guaranteed to rebuild a client's house if it was burned down. At the same time, he made considerable efforts to protect the house against fire, so that he would not have to pay to have it rebuilt. He organized bodies of fire fighters to turn out immediately on the occasion of a fire and he marked those houses that were insured with 'Barbon's Fire Office' with a lead badge – later, it was made of copper – depicting a Phoenix rising from the flames. When they arrived at a fire, the fire fighters would immediately check whether or not it had a badge on the door. If it did not, they would probably leave it to burn or, at the most, make only half-hearted attempts to put out the fire. If the house did have a badge, then they would do their utmost to save it and to extinguish the fire. Rebuilding was, in fact, no problem for Barbon. He had made a great deal of money rebuilding houses in the City of London after the Great Fire and then letting the houses out to those citizens whose own houses had been lost in the fire, together with most of their possessions and wealth, and who could not, in consequence, afford to rebuild their homes themselves.

Others quickly imitated Barbon, despite his attempts to claim exclusive rights on his insurance idea, and other firemarks soon appeared on insured houses. Among the earliest were the Hand-in-Hand and the Sun. Rivalry between fire fighters was intense and each would often leave houses insured by another company to burn. The wiser companies realized the advantages of advertising their services, however, and might try to save an uninsured house in the hope that the owner would be so impressed by their efforts and so frightened by the disaster that he would subsequently take out a policy with the office that had gone to his assistance rather than with one of its rivals. Needless to say, competition soon led to brawls between the fire fighters and, while they fought each other for the right to fight a fire, the house itself burnt down!

Engines were also improved. A typical pump of the period after the Great Fire was the Keeling pump of the 1670s. A pump of this type still exists and was used in Dunstable, England, until the nineteenth century. The four-foot by three-foot wooden tub which contained the pump was set on a wooden frame supported on wheels. Two handles stretched out at either end to raise and lower the piston in the cylinder. The tub had to be filled by bucket chain.

A similar engine was used in Belgium until the First World War. The Belgian machine was only slightly more sophisticated in that it had metal handles and a long metal nozzle which could be swivelled in any direction in order to aim the jet of water at the fire. But this engine, like the Dunstable one, consisted basically of a wooden barrel into which water had to be poured from buckets.

A more important development in Holland had far-reaching effects. In 1673, Jan van der Heyden introduced the fire hose, which enabled firemen to get right in close to the fire and direct their water jet at the source of the flames instead of spraying water from a distance over the top of the flames. Flexible hose was attached to the inlet of the pump, which at first was fed from a canvas trough propped up on a wooden stand but which was later fed direct from any one of Amsterdam's numerous canals. These pumps were not introduced into England until 15 years later. The hose itself was made of leather. In 1690, Jan van der Heyden published a book on fire engines in which there was an illustration of the old and the new types of fire engine, with and without hoses. The old style still retained the solid, swivelling spout which could only be used at a distance from the fire. The new type looked forward to a new interest in fire engines that saw many developments in the coming century.

Pumping by Hand

Richard Newsham took out a patent on a fire-engine pump that he invented in 1721 and claimed that he was the discoverer of the air vessel. We have already seen that there were other claimants to that discovery, some a long, long way before the eighteenth century. But although Newsham was not in reality the first, his engine did have many improved features that made it a landmark in fire-engine history and for a long time it was the model for all other manufacturers. By the 1730s, Newsham had perfected his engine, and we can still see examples of his machines in museums today.

A typical Newsham of the 1730s had handles on either side of the appliance instead of at the ends as in earlier engines. Such appliances were often known as side-pumpers. The advantage of the side-pumper was that more men could be fitted in along the pump handle, which in turn provided greater power. Even more power was provided by a set of foot treadles on top of the engine, where still more men could work the pump as they balanced themselves against the hand rails that were fitted along the top of the appliance. These men applied pressure with their feet to the descending treadle as it was worked by the men on the ground. There was, however, one fundamental disadvantage to the side-pump method: the appliance became unstable as

Left: George Washington presented this first manual to the Friendship Fire Company, founded in 1774.

the pump was rocked from side to side with the heaving of the men at the handles, and sometimes it might even tip over.

Two single-acting pump barrels were placed in a tank that formed the body of the appliance and the tank was filled by a chain of people bringing water in buckets from the nearest available source. The suction inlet was, in fact, fitted with a two-way cock so that the pumps could draw water either from the tank or from a suction hose. Delivery was by a pipe connected by a swivel joint or gooseneck with the outlet of the pump and fitted to the top of the appliance.

A contemporary book entitled *A Universal System of Water and Works* by Stephen Switzer quotes a circular of the period:

'Richard Newsham, of Cloth Fair, London, engineer, makes the most useful, substantial and convenient engines for quenching fires, which carries continual streams with great force. He hath play'd several of them before his majesty and the nobility of St James's with so general an approbation that the largest was at the same time ordered for the use of that royal palace . . . . There is conveniency for about 20 men to apply their full strength, and yet reserve both ends of the cistern clear from incumbrance, that others at the same time may be pouring in water which drains through large copper strainers . . . . As to the treadles on which the men work with the feet, there is no method so powerful with the like velocity and quickness, and more natural and safe for the men. Great attempts have been made to exceed, but none yet could equal this sort; the fifth size of which hath play'd above the grasshopper upon the Royal Exchange, which is upwards of 55 yards high, and this in the presence of many thousand spectators.'

Newsham had his rivals, however. One of these was Fowke, of Nightingale Lane, Wapping, who offered an engine with a double jet: 'Constant stream'd engines for extinguishing fires, the larger sizes play two streams at once, being the first and only of their kind, and does the office of two engines . . . . The four larger sizes run on wheels, and the other two carried by two men like a chair.' Newsham had his own reactionary views on the subject of a double jet: 'The playing two streams at once do neither issue a greater quantity of water, nor is it new, or so useful, there having been of the like sort at the steel yard and other places 30 or 40 years; and the water being divided the distance and pace are accordingly lessen'd thereby.'

Clearly Newsham was basking in contemporary praise, which said of him that by his invention, 'he gave a nobler present to his country than if he had added provinces to Great Britain'. But Mr Fowke was not to be put down. He commented on a trial that was held between his own engine and one belonging to his rival:

'Upon that occasion Mr Newsham attempted to show them that he could also play two streams at once, but his performance was so trifling, and done in so preposterous a manner, that he only got shame, but no credit thereby; nor does he know to this day the reason why his engine could not play half so far as Mr Fowke's play'd at Ratcliff. Wherefore by reason of such his ignorance he lately has endeavour'd to persuade the publick that the way of playing with double streams does not answer. Those gentlemen were otherwise satisfied that Mr Fowke's double stream had a much greater and more forcible effect, and that his engines might be worked by any strangers whatsoever. And they were as well satisfied that Mr Newsham's men (who were used to and instructed in dancing and treading in due time upon his engines) could only show them in the greatest perfection; and that by strangers it was utterly impossible to be effected. And therefore they rightly concluded that Mr Newsham was conscious to himself that strangers would tread confusedly and not keep time with the pumpers, and consequently would destroy each other's powers, and do more hurt than good, or else he would not have refused strangers to play his engine.'

America had already imported pumps from England at the beginning of the eighteenth century. In 1707, two pumps were brought over to Boston from England; in 1711, three more pumps went over to Boston to help to combat the high fire risk. In Europe, further developments occurred. Maundell and Grey joined the growing number of British manufacturers in 1712. Eight years later, a German called Leupold employed an air vessel in a very small appliance weighing only 16 pounds, with one cylinder, which was contained in a strong copper box. In 1725, an appliance without an air vessel was made at Strasbourg; this engine had pump handles only at one end. Much later, an interesting engine was produced at Ypres, mounted on a sledge and provided with an air vessel. This variation had pumps worked in a rowing motion, whereby the men took hold of pins projecting from horizontal bars and rowed the bars backwards and forwards. At the time this was considered extremely efficient, but it does not appear to have been widely used, so perhaps its brilliance was rather more theoretical than practical.

In 1774 it was the turn of John Blanch, who produced an engine that jetted water up to 150 feet high. In 1792, Charles Simpkin patented an improvement 'which consisted principally in the employment of separate valve chambers for containing the valves, instead of placing them within the cylinders and air vessels as was done previously'. These metal

Right: Benjamin Franklin, wearing the hat of the Union Fire Company, with the fire house to the right.

valves (not leather, as before) were placed so that they hung perpendicularly over the openings. This ensured that they opened and closed easily and without stress; their position also kept dirt out and made it less likely that they would be knocked about.

Over the years various improvements were made to Newsham's original design. Metal valves replaced the old leather ones, as they did in Simpkin's engine; treadles and chains were no longer used after a while; the wooden structure of the carriage was largely replaced by metal for additional strength; the length of both the carriage itself and the levers was increased so that more men could be placed along the levers to provide greater pumping power; the levers themselves were designed so that they could be folded up into a shorter length to make the whole engine less unwieldy when travelling to the fire; provision was made for the engine to be drawn to the fire by horses instead of by men; springs were added, so that less damage was done to the engine on the rough, pot-holed roads; more and more equipment, such as axes, hooks and ladders, was fitted to the engine; and the actual mechanics of the engine itself were improved so that the water could flow more freely and provide a smoother and more powerful jet.

Improvements in Europe were more than matched by progress on the other side of the Atlantic. In 1730, Philadelphia ordered two Newshams and one

local Nichols. In the ensuing decade, Newshams appeared in New York and Boston. In New York, local competition came from Zachary Greyaal, who tried to promote the use of a large water barrel which contained a gunpowder charge; his hope was that the charge would explode when the barrel was rolled into the fire, thus releasing the water to flood the fire. In both Philadelphia and New York, the fourth and sixth sizes of Newshams were purchased: the fourth could throw a jet of 90 gallons per minute more than 100 feet; the sixth could throw a jet of 170gpm about 120 feet. In neither engine did the wheels pivot: the appliances had to be lifted up in order for them to turn corners – a marked disadvantage in narrow city streets.

Benjamin Franklin was the man behind the Philadelphian interest in fire fighting and it was he who was responsible for starting the first volunteer fire company in America, the Union Fire Company, in 1736. This was the pioneer of the organized groups of volunteer firemen that were so important an aspect of American fire fighting for the next century. Forty years later, an even more widely known name became interested in fire fighting and the volunteer companies: George Washington himself was an enthusiastic fire fighter and followed Franklin's lead in trying to encourage fire prevention methods as well as up-to-date methods of fighting fires.

Before the American Revolution, however, there were few home-grown appliances that could com-

pete with the European models, in particular those from Germany and Holland, as well as the Newshams and Ragg and Nuttals in England. In 1743, Thomas Lote produced an end-pumper in New York, which became affectionately known as Old Brass Backs because of the amount of brass used to cover the pumping mechanism. In 1768, Richard Mason produced a similar end-pumper in Philadelphia. This required only six men to pump the smallest engine of the various classes, which could pump a jet up to 80 feet, and required 14 men to pump the most powerful engine, which could throw a jet to 120 feet. This, at last, was reasonable competition for the Newsham engines.

America was catching up and overtaking in other ways also. In 1752 Benjamin Franklin started up the first successful fire insurance company in the land: it became known as the Hand-in-Hand Company (a name already established in England) and its mark was the sign of four hands linked together in friendship and aid. Its real name was the Philadelphia Contributionship for the Insurance of Houses from Loss by Fire; it was hardly surprising,

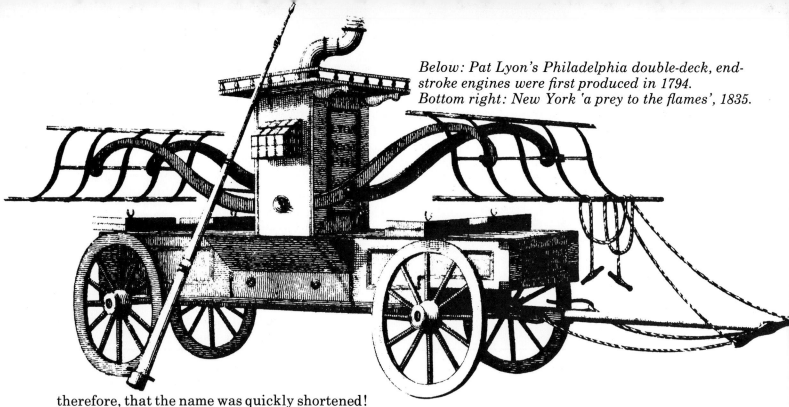

Below: Pat Lyon's Philadelphia double-deck, end-stroke engines were first produced in 1794.
Bottom right: New York 'a prey to the flames', 1835.

therefore, that the name was quickly shortened!

The war years of the Revolution were a bitter experience for the Americans as well as for the British soldiers. In 1776, New York itself was on fire immediately following Washington's withdrawal from the city. It was impossible to sound the alarm because Washington's men had taken all the bells in order to melt them down for munitions. More and more fires were started all over the city by incendiaries with firebrands, and the fire engines were sabotaged so that the British could not use them. Five hundred buildings were destroyed and nearly one-quarter of the city was laid waste in New York's first great fire. Sailors were brought ashore from ships in the port to help fight the fire, while all those people suspected of firing houses were cruelly lynched on the spot.

After the Revolution, the Americans came back to their towns to find many of their old engines broken and unusable. Their national pride and their memories of the dangers of fire during the war years, as well as their traditions of community service, encouraged them to rebuild their engines and themselves become volunteers. The volunteer forces increased enormously and, with the renewal of interest in fire fighting, both the standard of the forces and the engines they manned greatly improved. New designers emerged and many new names became quickly famous. Among these were James Smith and William Hunneman. James Smith's renown was based on his use of the gooseneck, the swivel joint that enabled the jet pipe to be moved in a complete circle without having to reposition the entire appliance. This was not a new invention, as we have already seen, but it had not been so consistently applied before to American engines. Furthermore, Smith's gooseneck engines were fitted with a fifth wheel under the chassis that pivoted the front axle so that the appliance could turn the

corners without having to be lifted. They also carried anything up to 200 feet of leather hose which could be added to still further by extra hose brought by members of the company in rolls over their shoulders.

William Hunneman built more than 700 appliances from 1792 onwards. He worked in Boston. His Fire King – a popular name for fire engines on both sides of the Atlantic – had smaller front wheels than those at the back to enable it to turn corners more easily. There was also Patrick Lyon, who pioneered the use of horse-drawn appliances. Before 1800, most fire engines had been hauled by hand. Not surprisingly, the volunteers, who prided themselves on their competitive spirit and the speed with which they could race to fires, were reluctant to accept the arrival of the horse-drawn vehicle and made their protests loudly heard.

Innovations in design were always greeted with scepticism, but the curious were nonetheless interested in having a closer look. Coffee-mill and cider-mill pumps were eagerly tried but were not widely successful. The coffee-mill pump had a handle at the side which wound a rotary wheel within the pump with teeth that scooped out the water down a pipe to produce a jet. In the cider-mill pump, the men ran round a windlass which had a similar rotary action within the pump. In 1800, America's first fireboat was launched in New York; this was a coffee-mill pump and the boat was rowed to the fire by volunteers.

Shortly afterwards, Reuben Haines introduced the hose wagon, which caught on at once with the volunteers, for whom it quickly became another chance for displaying their prowess and for decorating with glamorous pictures to embellish their image. Early hose wagons took up to about

600 feet of hose. At the same time, Sellers and Pennock started to use metal rivets to secure the seams of the hose instead of the primitive method of sewing that had been used until then. To make their hose, Sellers and Pennock used excellent quality cowhide, lengths of which they folded over to form a tube, the joints then being riveted. Light rubber hose did not appear until 1839, when it was introduced by Goodyear. The stronger hose made by Sellers and Pennock meant that engines could henceforward suction water more easily; this in turn led to improved engines providing a faster flow for quicker extinguishment of serious fires. It also heralded the end of the bucket chains. When water had to be transported some distance to a fire, several pumpers would be set up in line so that one stood at the place where the water was and suctioned from the water source, pumping the water on to the next engine in line, which in turn pumped the water on to the next, and so on to the engine at the end of the line which played its stream of water on to the fire itself. It was recorded at one fire that as many as 25 engines were set up in line over a distance of well over a mile between the source of the water and the fire.

New York suffered its worst fire in 1835, on a freezing cold December day. Losses amounted to $20 million and the whole of the Wall Street area was destroyed. It was New York's answer to London's Great Fire of 1666. A broken gas pipe set it going after the fire companies were already depleted by a run of previous fires. The weather was bitter, the fire risk was high, many appliances were out of use, hydrants and wells were frozen solid, and the wind, at gale force, was blowing the fire from house to house across the rooftops. Iron-closed shutters, intended for protection from rioters, held the fire within the buildings until suddenly it burst out with incredible ferocity, lifting the tin- and copper-bound roofs and bringing them crashing down upon the collapsing timbers of the main walls and beams. Crowds gathered to stare and get in the way of the firemen and shout encouragement as looters rushed out of the buildings with expensive rolls of cloth under their arms.

It was not long before the flames reached the blocks of buildings along the East River. Then an oil store blew up in a marvellous yet appalling spectacle. Four hundred volunteers from Philadelphia had to drag their appliances much of the way because the railroad on which they rushed their machines in answer to the call for help did not run close enough. In the end, the fire burned itself out. Ten thousand people were left jobless and New York was bankrupt. Not for the last time, Congress refused to bail the city out!

In the same year, Button and Company were building their piano-style engines, which earned their name from the large water boxes which looked very much like piano boxes. In 1840, John Agnew built his double-decker, which became

Philadelphia's most powerful and one of their most widely used pumpers. Double-deckers were pumped by two rows of men, one group standing on the ground and the others on the appliance itself. In one of the largest of these engines, 48 pumpers could jet a stream of water 180 feet horizontally: this was the Southwark Engine Company No. 38. Another style was the 'haywagon' produced by Waterman. This engine had double sets of handles on either side which could be folded up to give the impression of a haywagon with high-posted sides. The haywagon was more powerful than the Agnew, but it was also extremely heavy, a disadvantage that earned it the name of Mankiller. John Rodgers also built double-deckers in the 1850s, while James Smith continued to produce his rather more elegant side-pumpers.

There were also the curiously named 'squirrel tails' and 'Shanghais', both of which were still being built when steam pumpers were beginning to come into their own. It was Button who came up with the squirrel tail: the suction hose of this engine curved back over the rig when not in use and it was this curving 'tail' that gave the engine its nickname. Shanghais were so-named because of their pagoda-like decking: these were used in New York and were extremely powerful. Hand pumps continued to be used in America until the first decade of the twentieth century.

In Europe, similar improvements were being made. In 1820, William Baddeley invented the portable canvas dam or cistern, in which clean water could be collected from a fire plug so that it could be drawn off for the pump without danger from the stones, grit and dirt which were usually found in fire-plug mains. Hadley and Simpkin, who were the predecessors of the famous Merryweather firm, arranged handles on their engines that could be folded up when the appliance was on its way to a fire so that their great length would not get in the way. The same firm also applied springs to the carriage of the engine so that it would not be

Left: The pumps were helpless to save the Crystal Palace as this Currier and Ives print shows.
Right: Button and Blake hand-pumper, made in Waterford, New York, in 1863, and restored.
Below: Philadelphia-style end-stroke hand pump, proudly decorated and made between 1840–1850.

Above: Side-pumper of 1842. This is Henry Waterman's 'hay wagon', dubbed the 'Man Killer'.

jolted so fiercely on the roadway. John Cooper produced a rotary manual in 1827, which was set in competition with New York fire engines with the result that 12 men working on the rotary engine were able to throw more water than 36 men on two of the other engines. The trial was held at Boston, so there could be no accusations of British favouritism. In 1835, Tilley produced a small engine made entirely of metal. Tilley also provided designs for the construction by Merryweather of a floating fire engine for the Russian Government in 1840, to be placed at St Petersburg. This fireboat had an iron hull 60 feet long and 16 feet in the beam. The pumps were worked by four cranked handles 12 feet long and with radii of 18 inches. As an additional asset, the hull of the boat was double-ended so that it could go in either direction without having to waste time turning round. The pump could be worked by anything up to 50 men.

An engine made by William Roberts of Millwall experimented with telescopic handles, so that when fully extended the handles could be worked by more than 50 men, but when working in a confined space the handles could be contracted and then worked by only eight men. This engine was known as True Blue. Roberts also manufactured a small manual which could easily be disconnected from its carriage, to be used as a portable pump on shipboard.

Towards the end of the century, horse-drawn engines became increasingly common and the firm of Merryweather was among the leading manufacturers to introduce them. The Merryweather horse-drawn Paxton was a light fire engine highly suitable for country use where a heavier engine would quickly have become bogged down in the mud. Paxtons might be handled by 22 pumpers and would deliver 100gpm to a vertical height of 120 feet. Another Merryweather product was the London Brigade machine. (Shand Mason had a London Brigade steamer many years later; as with

Fire King, names became popular and were used again and again.) The Merryweather London Brigade was drawn by two horses and had a strong hand brake to hold it on the downward slope. The engine was mounted on springs and the men sat on top of the body, holding on to the handrail. There were two vertical single-acting pumps; these were driven by links from the horizontal shaft which was connected with the handles by levers outside the frame of the engine. Suction could be taken in through a length of hose or received directly from the hydrant. Merryweather also produced a small Factory fire engine in 1898, which was drawn by hand and could be worked by 12 men. One of these was used in Dorchester prison. Machines like these, simple and cheap to run, were still a feature of the fire-fighting scene even after not only steam but petrol, too, had appeared.

There were plenty of trials of manual fire engines and the results of a few of these give us some idea of the competitive qualities of the manuals commonly in use in the nineteenth century. Two machines, both manned by sturdy Guardsmen, competed in 1850, for example, and achieved relatively balanced results. One made 156 strokes in three minutes and in that time delivered 288 gallons thrown to a height of 117 feet; the other made 118 strokes in three minutes and in that time delivered 297 gallons thrown to a height of 130 feet. The first produced 96gpm and the second produced 99gpm.

At the Great Exhibition of 1851 in London, Merryweather entered a large and a small engine, Shand Mason entered a large engine, Perry of Canada entered one machine and Letestu of Paris entered two engines. During the tests it was noted that sometimes the pumps were linked together to form a chain from the source of water to the scene of the supposed fire. In one case, Merry-weather's small engine pumped by ten men fed the small French engine, also pumped by ten men, and was able to pump just about fast enough to keep up with the French engine. However, when both French engines, both pumped by ten men, were used to feed Merryweather's large engine pumped by 20 men, the combined efforts of the French engines were too much for the Merryweather: they supplied 30 gallons more in two minutes than the Merryweather could throw off. The Merryweather also suffered defeat at the hands of the large Shand Mason engine, also worked by 20 men: the Shand Mason produced 22 gallons more than the Merry-weather could cope with in three minutes. This swamping of the next engine in line was not unusual and, of course, led to immense rivalry between the pumpers of each engine. Those nearest the source of the water tried their hardest to pump more water than the next engine in line could itself pump on, and there would be cries of triumph when the second engine began spilling the water it was receiving. In the same way, the second engine would try to outpump the first engine, so that they could accuse them in turn of being slow and lazy and not supplying enough water.

Of the engines in this particular contest, the largest was Perry's with a pumping crew of 30 men, who delivered 188 gpm to a height of 132 feet. Their average gpm output per man was also higher than the other engines: it worked out at 6.27 gpm per man.

In 1855, 23 manuals were tried out against each other at the Paris Exhibition. The trials were taken very seriously by those who went, but apparently not so seriously by those who were putting the engines to trial, as one magazine report recorded: the trials were 'if possible worse managed than those of the English Exhibition of 1851 – the absurd and ineffectual attempt to systematize the experiments of that occasion being entirely omitted in Paris. Little account was taken, in the French experiments, of the various elements upon which the superiority must necessarily depend. Even the distance to which the water was projected was only ascertained *by stepping!* The exposed and shelving bank of the river [Seine] was but ill adapted for the trial and a maliciously high wind blowing at the time sadly marred the fair proportions of the performance, the best engines being necessarily the greatest sufferers.'

An American competition allowed three classes: 60 men for the first class, 50 men for the second and 40 for the third. But it was noted that only about three-quarters of these men could in fact find space at the handles at any one time. In the first class, a Brooklyn engine achieved a height of 110 feet and a distance of 206 feet; in the second class, one New York engine reached a height of 90 feet, but this engine achieved a lesser distance than another New York engine in the same class which reached a

distance of 164 feet; in the third class yet another New York engine reached a height of 84 feet (greater than the height achieved by one of the second-class New York engines) and a distance of 189 feet (greater than the distance achieved by any of the second-class New Yorkers). Quite sensibly, two of the best engines – including one that had thrown a jet of water 240 feet when manned by 44 men – were kept out of the trial and remained on duty in case of real emergency.

In sum, these trials only showed what could be done under 'almost' ideal conditions. Numbers of strokes varied between 35 and 65, manpower ranged between ten men and more than 40 men, gallons per minute ranged from about 40 to about 200, and distances thrown ranged from less than 100 to more than 200 feet. These figures were for tested periods of time ranging from one minute to three minutes, depending largely upon the energy with which the handlers pumped. It can readily be seen how at a real fire such rates could not be maintained for several *hours*, and in any case the pumpers would have to be changed every few minutes to recover. Constantly demanding refills of beer as they pumped, it was hardly surprising that the pumpers were a costly item and that those who had to pay them at the end of the day were prepared to welcome the steam pumps as a long-run economy.

In his book *Fires, Fire Engines and Fire Brigades*, published in 1866, Charles Young warned of the disparity between the achievements of engines at trials and in reality:

'A very absurd practice has obtained a firm hold with the builders of fire engines; namely that of estimating their engines as throwing off so many gallons of water per minute to a height of so many feet, which it will be found in actual regular work they never do really perform. For instance an engine is made and taken out for trial, manned with a lot of sturdy fellows, possibly some of the Guards, and these by straining every nerve, and getting possibly 60 strokes or more for a space of half a minute or so throw a few gallons of water to a very considerable height . . . . Woe be to the unfortunate purchaser who estimating the power of such an engine from these statements provides himself with one as he thinks calculated to meet his requirements. He will soon find himself thoroughly undeceived the first time its services are required for a reasonable spell of work.'

Young also had some things to say about some of the important factors in the construction of manual fire engines in the middle of the nineteenth century. There was, for a start, the type of wood that was to be used: various manufacturers preferred various types for various purposes. Hard oak might be used for the bottom of the body or cistern; best Baltic fir might be used for the sides of the cistern, or else Honduras mahogany or walnut; on the other hand walnut might be used throughout the engine; lever stays would probably be of oak, although ash was also considered. Above all, it was important, noted Young, not only in the choice of wood but also in the preparation of hose, rivets, joints and all materials, that 'no makeshifts, except under the most unavoidable circumstances should be allowed to exist for one moment in connection with fire extinguishing machinery'.

He explained the importance of his warning: 'It will frequently happen that long immunity from the ravages of fire will cause a diminution in the attention bestowed on the fire engine and appliances provided for its suppression; and they will consequently be certain to be out of order, or have some portion of the appliance deranged when wanted; and then as a rule it will have been found far better to have no engines at all than to have placed reliance in the means *supposed* to be efficient, and when taken out *found* to be all but useless.'

The machines, however, were only a part of the story of the manual pumps; the age of the hand pump was characterized equally by the nature of the men who worked at the handles – their rivalries, their roughness, their heroism, their parades and their pride. They were, often enough, the pride, too, of the local community, and the young men were the admiration of many of the local girls. Boys and youngsters ran beside the pumpers, hoping in time to be able to join their heroes at the handles and become part of what, in effect, was the most select club in town.

There were rivalries in abundance and plenty of foul play between the various brigades within a town. Indeed, in many cases the brigades became a force to be reckoned with in local politics and helped support politicians who favoured them, or terrorized those who wished to vote for their rivals. Each brigade would seek to reach a fire first, out of a sense of pride and to advertise its prowess. Brawling between the brigades was common, as much in America as in England, as we can tell by some of the rules of behaviour that were laid down by city councils once the rioting got out of control. In England, there were heavy fines:

'Any fireman challenging another to a fight: two shillings and sixpence.

Any fireman striking another: five shillings.

Any fireman throwing water or firebrands over another: two shillings and sixpence for each offence.

Any fireman cursing or swearing at another: threepence for each oath.'

On the other side of the Atlantic, we know that the District of Columbia had trouble as late as 1856. The following is an extract from the rules laid down in that city:

'Each company is required to be cautious in the admission of its members and prompt in the expulsion of the disorderly.

Above: This horse-drawn manual, made by Merryweather in 1866, was known as the Paxton.

In returning from fires, or alarms of fires, the companies are requested to move with order and moderation, and should they meet another company, each is requested to take the right of the street.

No company, unless prepared with its apparatus, to use the water of a pump, fire plug or other water source, shall retain it if demanded by a company thus prepared, Provided, however, that no company thus fully prepared shall attempt the violent possession of such pump, fire plug or other water source.

The use of intoxicating liquors among the fire companies comprising the fire department is required to be entirely discontinued.

However valuable as a fireman a member may be, if he is habitually turbulent and quarrelsome, it were better to dispense with his services as a fireman, than his continuance shall be the cause of constant broils and disturbances, and it is hereby recommended that every such member shall be stricken from the rolls of every company.

No company shall permit its apparatus to be run within any of the streets, lanes or alleys of the city, except when going to a fire or an alarm of fire.

The officers of the different companies are earnestly requested to accompany their apparatus on all alarms or fires, and on their return to the engine house; and to promptly check the first indications of a quarrel. In all cases when a riot or disturbance takes place, the company with whom it commences shall be held responsible.'

That gives some idea of the type of behaviour that was common. Furthermore, there were all manner of ruses adopted in order to be first at a fire. Racing to the fire, brigades would attempt to overtake each other or run each other into the gutter or into a wall. Arriving at the fire, they would fight for the use of the available hydrants; sometimes they would send men ahead to lay claim to the hydrants and guard them against rival brigades until their own engine

the handles came crashing down; legs might be trapped by the foot treadles on the larger machines; fingers certainly would get pinched and occasionally cut off.

There is an excellent description of firemen racing to a fire in a poem in 'Rejected Addresses' by Horace and James Smith, published at the beginning of the nineteenth century.

The summon'd firemen woke at call,
And hied them to their stations all:
Starting from short and broken snooze,
Each sought his pond'rous hobnailed shoes;
But first his worsted hosen plied,
Plush breeches next, in crimson dyed,
His nether bulk embraced;
Then jacket thick, of red or blue,
Whose massy shoulder gave to view
The badge of each respective crew,
In tin or copper traced.
The engines thunder'd through the street,
Fire-hook, pipe, bucket, all complete,
And torches glared, and clattering feet
Along the pavement paced . . .
The Hand-in-Hand the race begun,
Then came the Phoenix and the Sun,
Th' Exchange, where old insurers run,
The Eagle, where the new;'

The smart outfits served to attract volunteers to the brigades and they also served to advertise the brigades to the public, as did the splendidly painted engines and the pomp and ceremony of the parades in which the firemen loved to indulge on holidays. But the outfits were not always very practical. Fancy buckled shoes were often worn instead of decent boots and these provided little protection from water and heat and rough, broken masonry, as the watermen – for in London many of the early fire fighters were watermen from the barges and ferries on the River Thames – made clear when they put in a petition for decent boots. In the 1760s, they begged the governors of the London Assurance:

'That amongst all the dangers and hardships to which your poor petitioners are exposed in the discharge of their duties of their office, there is none more sensibly felt than those which they experience from the want of boots, as their leggs are frequently torn with nails, barrs of iron, and such kind of rubbish as fires occasion.

That the late dreadful fire at Shadwell particularly evinced the great necessity of boots, as several of your poor petitioners were up to their knees in water, hot from the waterworks, and instantly after plunged in cold water, by which deplorable case great numbers of your petitioners lives were endangered, by the coughs and colds which they caught, which calamity, your petitioners presume to suppose, might be in future happily prevented were their leggs defended by boots.'

There were, however, inefficiencies in the system – whatever the gallantry of the men – as Mr Baddeley,

arrived. Inevitably, fights ensued. One of the most terrible disgraces that could be suffered by a brigade was to be overtaken, so to avoid this fire leaders would suddenly call their men to a halt, just when they feared they were about to be passed, claiming that they had to make some necessary repair to their engine before it could go on.

The companies fitted bells and gongs to their engines to clear the way, and some companies brought their own water with them on the appliance so that they could start pumping as soon as they arrived. Other companies claimed that this was cheating and so the practice was no longer allowed – whatever its evident advantage to fire-fighting techniques! The men began to bunk in the fire house for the night so that they would be quicker away in the case of a night alarm – and most of the alarms seemed to be at night. When some companies started to use horses to pull their engines, they were mocked and again the cry of 'cheat' was raised.

The men themselves had to be tough to work the pumps. As the rate of pumping rose from 80 or 90 strokes a minute to 130, 150 and even as high as 170, the length of time for which a man could keep pumping grew shorter and shorter. Volunteers had to be called in from the crowd who stood around. Often there were injuries, as they tried to leap in to seize the handles without letting the rhythm of the pumping cease for one moment: arms would be broken as the designer, pointed out in 1838: 'In one short street (Lombard Street in the City of London) there are nine fire engines, and in the turnings out of it eight more, making in all seventeen within a few square yards; while there are many spots of at least three times the extent, and containing ten times the amount of property, without a single engine for their protection. Notwithstanding the proximity of the above engines to the Royal Exchange at the late fire, one of the smallest only was brought out and that one, as a matter of course, was not worked!'

And in the provinces, the fire engines were not always quite what might be hoped for, as described by Charles Dickens in the middle of the nineteenth century: 'We never saw a parish engine at a regular fire but once. It came up in gallant style – three miles and a half an hour at least; there was a plentiful supply of water, and it was first upon the spot. Bang went the pumps, the people cheered, the beadle perspired profusely but it was unfortunately discovered, just as they were going to put the fire out, that nobody understood the process by which the engine was filled with water, and that eighteen boys and a man had exhausted themselves in pumping for twenty minutes without producing the slightest effect.'

In other aspects of fire fighting, there was better progress. Sprinkler systems were invented, first by Carey in 1806, and then by Congreve in 1812; the Manby extinguisher appeared in 1816, with the gallant Captain pictured alone against the flames holding his small household fire extinguisher. In the 1830s, the Wivell fire escapes came on the market, with their fly ladders. These consisted of a main ladder, which could reach the second-floor window of a house and was mounted on a spring carriage with two large wheels; hinged to the top of this ladder was a fly or extension ladder which, by means of ropes, could be swung up to the third-floor window; an additional ladder with hooks could reach to the fourth floor. Beneath the main ladder there was a canvas trough or shoot, down which people could be slipped to the ground.

The image of Mr Wivell, mounted on his ladder, wielding an axe to gain entry at a closed window behind which the incarcerated family were being threatened with an agonizing death, made the fireman a gallant figure in Britain. In America, meanwhile, he was increasingly recognized by the characteristic helmet first designed by Henry Gratacap of New York.

Thrown from a window, it became the signal of a cry for help from a trapped fireman. By the middle of the nineteenth century, help was on its way in the form of more water to combat the blazes of destruction.

Far left: Drawing by Abraham Wivell of shoot escape and fire ladders invented by him, 1825–1837. Below: Hose reel and hand pump compete in this 1854 print, The Life of a Fireman – the Race.

The Age of Steam

It might be thought that the arrival of steam power would have been greeted with sighs of relief by the men who toiled at the hand pumps. Not at all: they looked on steam with suspicion, scorn and not a little fear, for it heralded also their loss of prestige, their beer money, their club-like groups; it brought a new age of professionals that was anathema to the courageous amateurs.

Braithwaite and Ericsson's steam engine of 1829 met with the most determined opposition, despite its initial successes at fires in the London area. This engine was the first ever to be powered by steam, but as a result of the opposition London had to wait more than 20 years before steam power came to be to be accepted in the city. In the interim, steam engines were employed in other cities in England, in America and in European countries.

John Braithwaite's steam engine had two cylinders and was capable of ten horsepower. The boiler was stoked from the rear of the engine with coke and at one of its first employments at the Argyll Rooms, in London's Soho district, three bushels of coke were consumed in five hours' pumping. The pump itself was at the front of the appliance, behind the driver's seat, and the two pistons of the engine were connected directly to the two plungers of the pump. The pump had a large spherical air vessel above it, and below

Left: 1875 hose carriage by Button & Son, beautifully restored, now at Salem, New York.

there were connections for suction and discharge hoses. The weight of the entire appliance was 45cwt and it was capable of throwing 150gpm to a height of 90 feet. Steam could be raised in 13 minutes.

On the occasion of the fire at the Argyll Rooms, the cold was so severe that the manual engines brought to the scene froze up completely and could not be used; it was only the Braithwaite steamer that kept going and was responsible for dousing the fire, even though the building was in fact totally burnt out. Despite this success, those who condemned the steamer persisted: they said that steam took too long to raise; they complained that it was too powerful for ordinary use, too heavy to travel fast enough to the scene of the fire, and that it needed more water than could be supplied to it in the streets of London; finally they moaned that, even should such amounts of water become available, the steamer would then spread so much water in the building that the water itself would do more

damage than the fire! There was no pleasing the critics. What would they have said to the 1000gpm and 2000gpm pumpers of today!

In the face of this opposition, Braithwaite built four more fire engines, none of which was used in London. The second was completed in 1831 and consisted of a single cylinder with a five-horsepower output; it was capable of a jet of water that could reach more than 100 feet. This engine served well in both France and Russia. The third engine, built the same year, was sent to Liverpool. It had two cylinders and had three pumps placed horizontally; there was gearing between the pistons and the pumps. The engine saw several years' successful service. Braithwaite's fourth engine was one of his most famous. This was the 'Comet', which he built for the King of Prussia in 1832. It was 15 horsepower, with two pumps and two cylinders, and weighed a total of four tons; it had an output of approximately 300gpm. In trials at Paddington

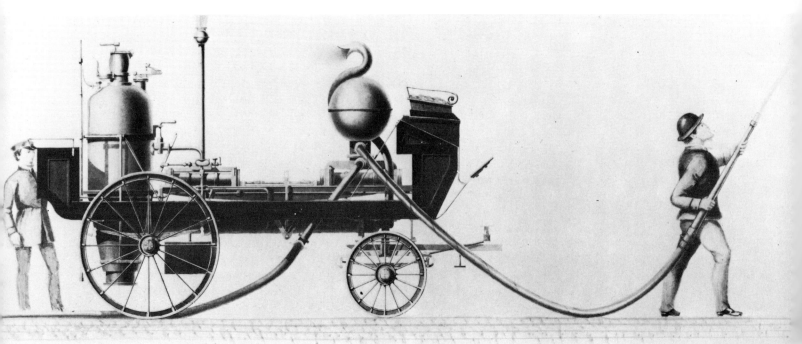

Canal before the engine was sent to Berlin, it was ready for action in about 15 minutes and was able to throw a jet of water vertically to 120 feet; at an angle of 45 degrees, water was thrown as far as 164 feet. The fifth and last engine was built the following year to an experimental design which, from the little we know about it, we must assume to have been rather less successful than the earlier engines.

In America, the steam engine was heralded by Paul Hodge in 1840. The time was ripe, for New York had recently suffered its major fire of 1835-6. The gooseneck hand-pumpers had proved totally inadequate to cope with the conflagration; their pipes had frozen in the cold and $20 million worth of damage had been suffered. Hodge's steamer was also the world's first self-propelled steam engine, so there was reason to hope that the future might be more secure with the inventive spirit of America in full swing. Closely resembling the early railway locomotives, the Hodge fire engine had large rear wheels which were jacked up off the ground when the pump was in operation and acted as fly wheels for the pump. The engine could also be drawn either by hand or by horses; it weighed approximately two tons. In claiming his fire engine to be the first in America, and in discounting claims made that John Ericsson had gone to America and himself constructed the first American engine, Hodge wrote: 'I would remark that I never saw a steam fire engine in my life up to this period.'

Other American designs quickly followed, while in Britain the steam fire engine remained undeveloped. In 1851, a self-propelled steam fire engine was designed by Mr Lay of Philadelphia. This had a rotary pump and was designed to throw 300–400gpm; its total weight was one and a half tons. Two years later, Mr Latta of Cincinnati constructed another self-propelled steamer. This ran on three wheels, having a single wheel at the front which could turn in any direction in order to facilitate the steering of the appliance. With two cylinders, the engine could produce enough power to throw up to six streams of water. Its best performance included a jet of 240 feet and up to 2000 barrels of water in one hour. At one fire, the engine was fully working within only five minutes of arrival and worked without stopping for eight hours.

Another Cincinnati steamer was made by Abel Shawk in 1855. This was able to raise 120lbs per square inch of pressure and was able to throw water 120 feet vertically and 172 feet horizontally. With four nozzles in operation, the pump was able to throw each approximately 100 feet. Three years

later, Poole and Hunt of Maryland began to manufacture steamers and produced approximately one for each year within the next seven or eight years – that is a measure of the amount of work that had to go into these appliances. With the emphasis on simplicity, however, Poole and Hunt made three chassis sizes with engines capable of raising steam pressures of 50–60lbs per square inch. The varying engines, each having 100 feet of hose but employing various sizes of nozzle, produced jets of 257, 240 and 235 feet.

Among other early manufacturers were Ettenger and Edmond of Richmond, Virginia; Reaney and Neafie of Philadelphia; Merrick and Sons of Philadelphia; and Lee and Larnard of New York. In 1859, trials in Philadelphia between steamers from these and other manufacturers produced jets varying between 100 and 200 feet and starting-up times varying between 11 and 18 minutes from cold. One of the most successful was the No 7 Baltimore built by Poole and Hunt, which achieved a starting-up time of 11 minutes dead and threw a jet 196 feet.

Trials were all the rage, both among steamers themselves and between steamers and manuals. By 1859, interest in steamers had rekindled in Britain. The next decade was the high point of the early trials and saw raging arguments over the merits and demerits of steam power. The sense of competition and pride rode high.

In 1858, Shand Mason produced their first land

53

STEVENS

steam engine. The only steamers in the period since Braithwaite had been constructed for use in fireboats. This was partly because the head of the London Fire Establishment, James Braidwood, was adamantly against the use of steam by his men. He claimed, with others, that the steamers would do damage with the amount of water they produced; more important, he believed that their extra power would discourage his men from getting in close to the fire, which was a vital part of their training and of their success under his direction.

The Shand Mason of 1858 weighed four tons and was drawn by three horses – the self-propelled engine did not appear in Britain for another four years. Despite the passage of time since Braithwaite's first engine, nearly 30 years before, one authority maintained that the Shand Mason was 'in point of efficiency, simplicity, durability of parts, weight and cost . . . in no respect superior to Mr Braithwaite's engine of 1829, while in some respects it was inferior to it'. The following year James Skelton constructed the first steam engine in Ireland; it weighed 22cwt and had a single cylinder.

Merryweather came up with their first land steamer in 1861. This was the Deluge which had a 30-horsepower engine and a single horizontal cylinder with a double-acting horizontal pump which worked directly off the piston rod of the steam cylinder: there was therefore no crankshaft, crank, fly-wheel or connecting rod. The Deluge achieved a

jet powerful enough to throw water clear over a chimney 140 feet high; later it achieved a jet 215 feet horizontally. The engine took part at the first trial of steam engines held at Hyde Park in London during the 1861 International Exhibition.

There was some confusion at these trials, which were originally intended to have been between Merryweather's Deluge and a small five-horse-power engine with a rotary pump made by Lee and Larnard of New York. Mr Lee himself had come over with his engine but refused to compete, on the grounds that the judges were not qualified to pass an opinion on the subject of steamers and that the tests in any case were of little point. So, instead, Shand Mason produced two of their steamers to compete with the Deluge in order not to disappoint the crowds who had gathered to see the fun. The machines were fairly matched but the committee concluded their judgements by noting that 'the results of these trials show that although something has been done towards making a really serviceable steam fire engine, still much remains to be accomplished. These results show that a decided advantage is obtained in working fire engines with steam as compared with manual power, but there appears on the part of all the makers a decided tendency to run their pumps too fast.'

Two years later, trials took place at the Crystal Palace, near London, and Merryweather had two new steamers competing, the Torrent and the

Sutherland. The Sutherland won first prize for the large steam engine class; it was immediately bought by the Admiralty to be used in their Devonport Dockyard, where it worked until 1905, when it was handed back to Merryweather. This engine was capable of maintaining a jet of water up to 170 feet high. It was a double-cylinder engine, whereas the Torrent had only a single cylinder and was placed in the smaller class at the trials.

In the same year, Shand Mason produced an early version of the vertical type of engine, which they named the London Brigade Vertical. Their appliance of 1863 was the first example of the short-stroke, high-speed type of engine to be used in a fire appliance. By 1876, the London Brigade Vertical had become a great deal more sophisticated. Carrying enough fuel to see a steamer through a long fire had often been a problem, and additional fuel had frequently to be brought in separately to feed the furnace. In the 1876 Vertical, however, there was a coal bunker under the fore carriage and conveniently near to the door of the furnace. This model had another advantage: there was a special fitting on the draw-pole whereby a horse that had fallen in its traces could quickly be released. In many cases, one fallen horse could bring disaster to the rest of the team and might easily overthrow the entire appliance. Racing to the fire, it was only too easy for a horse to slip, and many engines were thus

Above: The Deluge was the first steam fire engine manufactured by Merryweather & Sons, in 1861. That same year, it was entered for trials at Hyde Park, in London, for the 1861 International Exhibition.
Far left: Lee and Larnard's steam fire engine, from the Mechanic's Magazine, April 1862.
Below: The horse-drawn Sutherland, built by Merryweather & Sons in 1863, is believed to be the oldest steam fire engine still in existence.

brought to disaster; usually it would take some precious moments to disentangle a dead or injured horse.

At the 1863 trials at the Crystal Palace, there were ten entrants: seven were British, three American. The American engines, brought over especially for the occasion, were two built by the Amoskeag Company and one built by Lee and Larnard, who had apparently not been wholly deterred by the experience two years before. Perhaps Mr Lee had not been altogether unreasonable in his original complaints, however, because we learn that here again there was apparently no one on the committee which was judging the trials who had any idea of the working or manufacture of fire engines – perhaps this was sheer prejudice against the innovations they presented!

There was, in any case, an unfortunate incident at the beginning of the trials. Once again, it involved the machine belonging to Lee and Larnard, the Manhattan, which was in the hands of the London Fire Establishment. After the engine had been weighed and passed for the test, it set off for the area of the trial, but it took the wrong route – with horrifying results, as one observer recorded:

'Instead of the engine being taken by the easiest and nearest way to this point, it was taken round by the north tower, where there was a steep incline; but when it reached the top of the incline, and began to descend, the weight overpowered the men in charge of it, and the man of the London Fire Establishment, who had hold of the pole, was unable to guide it, and it ran away down the incline, and, at a point where the road curved, ran into a tree and capsized, smashing the engine and severely injuring the fireman before alluded to . . . . The force of the blow knocked off the forecarriage, broke one of the fly wheels, and cracked the other, turning the engine completely upside down, thus leaving the engine the night before the trials in a most crippled condition.'

Undeterred, the American delegation managed to get their machine ready for the trials, tested it in case it would blow up as a result of its crash and kill the onlookers, and then put up an extremely good show against the Sutherland. The degree of competition and the sense of indignation at the time between the American and the British manufacturers can be judged by the comment of the Americans present, who were said to have declared after the trial that, 'Had there been a fair trial and no "accident", the "Manhattan" would have been successful even against the rules, regulations and prejudices of the English committee.'

An engine named the Princess of Wales was also at the trials. This was the second steamer to be built by William Roberts of Millwall. It was not self-propelled. Robert's first engine, built in 1862, had been the first European self-propelled steamer: it was a three-wheeled machine, with a single wheel at the front, had a vertical boiler, weighed seven and three-quarter tons and had a maximum road speed of 18mph.

Recording the trials at the Crystal Palace, *The Times* concluded, in recounting the performance of the Sutherland: 'It threw a most magnificent column, and maintained the column steadily. This was the most massive jet thrown. This last effort was exceedingly beautiful to witness and brought the trials to a close. The engine maintained its speed with great regularity, the steam never exceeding 100lbs psi, and the firing was easy with nothing like forcing. The water pressure was between 85 and 95lbs on the square inch, fluctuating of course at each stroke of the engine, and the jet was perfectly solid and without air.'

However, there was another small episode to mar the day. Yet more indignation was aroused when a small Shand Mason steamer which had failed to draw water satisfactorily during the trials themselves was subsequently given a private trial, approved by the committee, at which it apparently 'worked well throughout' and was accordingly

awarded a prize. Charles Young protested loudly: 'How often it happens that a racehorse will run exceedingly well *by itself* but gets beaten when run against others, everyone knows; but it has never yet been heard of a beaten horse getting up a race by itself, in order to obtain the prize it could not win when running with the others. This is precisely the case as it occurred at these trials, but instead of horses there were engines; and, in spite of the protest of the other competitors, the prize was awarded to an engine which, in open competition at a given place on a given piece of work, had utterly failed in even starting.' As another spectator remarked, who did not in any case believe in steam fire engines, 'What would have been the consequence if, instead of being at a trial, it had happened at a fire, and an hour had been occupied in vainly attempting to get the engine to work.'

Still more virulent were the competitions between manuals and steamers and the ferocity of feeling

Left and below: Two views of a model of the Shand Mason London Brigade Vertical steamer of 1876.

Above: Roberts' self-propelled steamer of 1862.

between the manual pumpers and the exponents of steam power. In America the hostility between the volunteers on the one hand and the steamers that were going to make possible a salaried fire department of professionals was more bitter than anywhere else. With the volunteers and manual pumps, a salaried, organized fire department was quite out of the question, for no department could possibly have afforded to employ the number of men needed to handle the pumps. When, in 1852, Latta and Shawk built their first successful steamer, the volunteers jeered them and even threatened their lives. Five thousand dollars were appropriated to commission the partners to produce a steamer that would have six jets worked simultaneously, but while working on the commission Latta recorded: 'I sometimes feared that I shall never live to see this grand idea brought into the service of the world. The recent riots here show what a mob can do in our city. My steps are dogged. Spies are continually on my track. I am worried with all sorts of anonymous communications threatening me with all sorts of ills and evils unless I drop work on this engine and pronounce it a failure. I'll never give up! I'll build it and there are men enough in this city to see that it has a fair trial. When it is finished, it will be heard from at the first fire, and woe to those who stand in its way!'

The 'sham squirt', as it was called with derision by the volunteers, was tested against a Hunneman on New Year's Day, 1853. Answering the call to the test, the giant steamer crunched its way along the cobbles and puffed and smoked as it got up a head of steam. Meanwhile the volunteers quickly set their hand pump in action and were soon producing a fine jet of water that went 225 feet, with such pressure that two men were needed to hold the nozzle. But it did not take long for the volunteers to tire and come to a stop, while the steamer kept steadily pouring out its stream of water and then began to turn on its other five jets just to show what it could really do in case of emergency. The point was well made, and Cincinnati voted in a salaried fire department soon afterwards.

Comparative costs between manuals and steamers were carefully studied and publicized. In one set of comparisons, in London, it was noted that a steam engine exerting 30 horsepower would do the work of 150 strong men. In London the men on the manuals were then paid one shilling for the first hour and sixpence every subsequent hour. On a fire that might last nine hours, with men being changed whenever necessary, and supplies of beer almost unlimited, it can easily be seen that the costs could mount alarmingly. On the other hand it was maintained that a steam engine, with attendants and fuel, need only cost about 15 shillings for the same period. It was added by the proponents of steam that from four to five manuals were needed to produce the

same amount of water as one steamer. Of course, there was the initial cost of the steamer, which was too high for many small towns, but it could quickly pay its way.

In America, in the 1860s, the figures spoke equally clearly. A steam fire-engine company consisted of between seven and 14 men, of which number several would have other employment and turned out only in the event of a fire. Expenditure on, for example, an Amoskeag in the city of Manchester, New Hampshire, was approximately $860 annually. A first-class hand-engine company, consisting of something like 50 men, might well cost several hundred dollars a year while doing no more than a quarter of the amount of work.

Despite these figures and the tests that had been made, the manual was still virtually the only defence London had against the dreadful Tooley Street fire of June 1861. To support the manuals, there was a single Shand land steamer and a steam float from the river. It was in this fire that London's first highly successful and energetic fire chief – James Braidwood – was killed. The fire began in a ware-

Above: A steamer from the fire department of the town of Cincinnati, turning out to a fire, with excited onlookers.
Far left: Hunneman steamer from Boston, a common sight in the American 1870s.

Below: The Citizen's Gift was a worthy successor to Latta's first steamer for Cincinnati, Uncle Joe Ross.

house in which hemp was stored. The cause was spontaneous combustion: the iron fire doors that were supposed to contain the area had almost certainly not been closed. It was only seconds before the whole warehouse was blazing furiously, fuelled by other items in store, such as tallow, saltpetre, cotton, rice, sugar, tea and spices. Braidwood brought every available pump close in to the fire in his typical method of attack. The heat of the fire was so intense from the riverside warehouse that the woodwork of the fire float was blistered and the faces and hands of the firemen were scorched. Ten thousand casks of tallow went up in the blaze, and rats started to pour out of all the neighbouring buildings. As anxious about his men as about the fire itself, Braidwood set off down an alley, taking a short cut to go and see what was happening at the riverfront. A wall collapsed suddenly and he was buried and instantly killed.

Without their leader and with melting tallow flowing over the surface of the Thames, burning and keeping all boats at bay, the firemen were not able to bring the inferno under control for two days. Braidwood's body was not recovered until the day after that. His funeral was a grand gesture of respect: 'There was a concourse of people which had not been equalled since the interment of the Duke of Wellington', noted one magazine. But Londoners quickly regained their confidence when he was succeeded by the high-handed and aristocratic Captain Shaw –

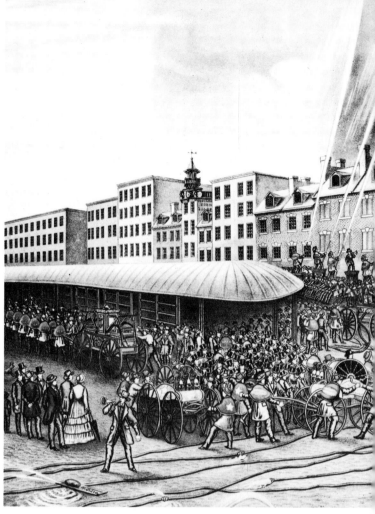

Above: This 'great engine contest' took place on Sunday evening, 7 July 1850, at Market and Fifth Streets in Philadelphia.
Right: Captain Sir Eyre Massey Shaw, organizer and chief of the London Fire Brigade from 1866.
Left: The Tooley Street fire of 1861, which broke out in a Thames-side warehouse in London and caused the death of James Braidwood, London fire chief.

in marked contrast to Braidwood, but nonetheless a worthy successor who continued to improve the fire service.

The 1860s saw further trials as steamers became increasingly popular. In 1864, there were trials at Rotterdam and Middleburg. In both places, a single-cylinder vertical Shand Mason competed with a double-cylinder horizontal Merryweather. As in other European towns, so in Middleburg, the town put up the expenses for the trials so that they could make a test on their home ground as to which of the currently available machines was best suited to their own particular needs. Manufacturers were only too happy to bring their machines across to Europe, hoping for new commissions. A year later, a competition was arranged in Cologne, where medals were apparently given not so much for winning as merely for taking part, in order to keep everyone happy. There were, in fact, cash prizes for the outright winners.

By this time, the number of steam engines being built had increased considerably. In 1863, a steam fire engine was built by Cowan of Greenwich for the Library of St Petersburg. The engine weighed four tons and was drawn by three horses; it produced a vertical jet of 170 feet and a horizontal jet of 210 feet. In Hanover, the same year, Egestorff built a single-cylinder steam engine which was a little less powerful than Cowan's. In 1864-5, Shand Mason built 17 fire engines: two of these were for the London Fire Engine Establishment, two for Lisbon, three for the Bombay and Baroda Railway Company, four for Russia, two for New Zealand, one for Austria, one for Poland, one for Denmark and one for Dublin.

In 1865, Roberts built his third engine to order for the Arsenal at Rio de Janeiro. This, the Excelsior, was very much the same as the Princess of Wales. Trials were held at Millwall before dispatching the Excelsior to its final destination. It was ten feet six

inches long, four feet eight inches wide, weighed 31cwt and had three wheels. In the same year, Moltrecht of Hamburg and Flaud of Paris produced small steam engines and Merryweather produced 11 steamers which went to many different places around the world: two for the Spanish government, one for Portsmouth dockyard, one to Liverpool Corporation, one for the French government, one for Amsterdam dockyard, one for Dublin, one for Manila, one for Redruth, a floating engine for the North Eastern Railway for use in Newcastle, and one 'waiting orders'.

Writing in 1865-6, Charles Young took an intense interest in the development of these engines and took great care to visit as many of their trials as he could. He also outlined the requirements in good steam engines, how they should be managed, their advantages over manuals, and how they should be constructed. Verifying his findings with engineers of the day, he outlined the following requirements:

'Quickness in raising steam, so as to be ready for work in the shortest possible time; simplicity of construction, so as to be easily worked by any person of ordinary intelligence; strength, so as to avoid break down from hard work at fires, or from travelling rapidly to them; durability, so that its working may not damage it and cause it to be frequently under repair; lightness, to enable it to be easily and rapidly drawn, with all its accompaniments, by a couple of ordinary horses; efficiency, to enable it to perform the greatest amount of work for its size and weight; and manageability, to ensure its useful employment when at a fire, and to enable its powers to be varied with ease and certainty, in accordance with the varying requirements of a fire, and thereby enable it to produce the greatest effect with the minimum of damage.'

Quick to defend the steamer against the manual, Young championed it in every way, adding that, since the steamer was under the control of one man, 'instead of being worked by some 30 or 40 excited roughs duly primed with unlimited beer', it was likely that the steamer could be controlled and adjusted much better to the variations between a small flow one minute and a great jet of water in the next minute, as the fluctuations of the fire necessitated.

As for the management of the steam engine, many of the directions he gave might equally apply to the fire engine of today, as they did also to the manual engines of the previous century. Young emphasized that 'the careful maintenance in working order of a steam fire engine, its thorough examination when it is in the engine house, and its judicious management whilst working, are essential to obtain the full

Left: Clapp and Jones were popular manufacturers of steam engines. This 1888 model was capable of 600gpm.

development of its powers, and to ensure its satisfactory performance'. He quoted James Braidwood, the victim of the Tooley Street fire: 'One of the great objects to be attended to at a fire is the safety and preservation of the fire engines; for if these give way while working, it is rarely possible to procure others without a fatal loss of time.'

Accidents could happen from any number of causes, some of which were similar to those suffered by the manuals. The suction hose could inadvertently draw up stones and grit. This would inevitably jam the valves or cause resistance in the pistons, which would eventually make them crack. A suction hose that was leaking at the joints could draw in air if not properly secured. Yet more disastrously, leaks could occur if the engine was started up too suddenly, with too much steam; the engine would quite probably break down from condensed water in the cylinders when starting up in this manner.

Different makes of engine had different requirements, but a fairly typical one at that time might have an engineer, a stoker, and two firemen to manage the hose. The engineer and the stoker were responsible for the everyday management of the fire engine. Both had important jobs. Among the generally accepted rules for preparing the engine for operation were those governing, for example, the laying and starting of the boiler fire. Very dry wood shavings were needed for this, as well as a few small knobs of coal and some pieces of small, well-tarred wood. If it was necessary for the fire to start as quickly as possible, 'about a pint or so of turpentine, camphine, or other inflammable stuff may be thrown in just before lighting up; but this will be found to cause an abominable smother, and should only be done where necessity can vindicate such a nuisance'.

The fire was not always lighted immediately after the alarm was sounded: sometimes lighting was delayed until the engine had already covered some of the distance to the fire. 'If the fire is within ten minutes' run of the station the fire should be lighted at once,' advised Young; 'if it is any greater distance, then it may be found better not to light up until within eight or ten minutes run of the scene of the fire. Most steam engines are capable of being set to work in from eight to ten minutes after lighting the fire, providing it has been properly laid, the right amount of water be in the boiler, and the engine maintained in proper condition.' Just how the boiler-man managed to light the fire while the steam engine was racing to the scene of the fire is difficult to imagine, but it was clearly true that, the longer the lighting of the fire was delayed, the safer. Secondary fires were often caused by sparks from the furnace of a steam engine; these sparks easily caught on any dry matter and set off fires which had to be attended by the engine immediately or when it had finished its work at the first fire.

There was another danger for the stoker to watch for, however, as Young warned: 'Sometimes it will happen, if the engine has to run a good distance over rough pavements or roads to get to the fire, that the jolting and bumping will have shaken and packed together the wood, shavings and coals into a tolerably compact mass, so as to cause the engine to be longer in getting to work from the fire not going off quickly.'

Regulation of the fire was, of course, important, for it was vital to maintain a proper supply of steam at the required working pressure and to maintain the water at its proper working level. Fuel had to be supplied regularly and 'so managed as not to let the pressure of the steam vary too much. The most favourable and proper time to run the water.' wrote Young, 'is when the steam is blowing off smartly and the fire is strong.' There were emergency measures in case the steam was generated faster than the engine required. In the first instance, the thing to do was to open the fire door wide; if this did not work, a handful or two of water should be thrown on the fire. Admiring the American steamers, Young said that he had seen their firemen 'keep both water and steam beautifully exact, the engine running at 200 revolutions and upwards per minute, and the steam at 150lbs psi, by simply opening or closing the fire door and regulating the quantity of feed, and this too without the slightest fuss or trouble'.

Cleaning the engine properly at the end of the day was one of the most important aspects of maintenance. Rust could rapidly corrode the metal and cause irreparable harm to the engine. Brasswork had to be kept immaculate, for it was as much the pride of the fireman as well as a spectacle for the citizens who watched the splendid machines careering through the streets to the scene of the fire. But the duties of the fireman, and his responsibilities in cleaning and working the machine, were not for the dainty, as Young pointed out: 'It is no use for anyone who is afraid of his delicate fingers, or is an extensive patroniser of kid gloves, to have anything to do with steam fire engines, there being far too much "reality" about them to associate comfortably with any "ideal" of this description; consequently, as the cleaning and keeping of such a machine in proper order must not be neglected, nor done by halves, it is most important and desirable that this should only be done by those who do not labour under such an affliction.'

An important distinction that was made in the types of fire engine was between those that were slow-running and those that were quick-running; there were also those in which the steam piston ran faster than the pumps. In the slow-running machine, the engine and pump had a long, steady stroke, and were run at from 50 to 100 strokes a minute; in the quick-running machine, the engine and pump had a short stroke and were generally run at between 150 and 200 strokes a minute. The manufacturers of

Above: Steam and muscle compete in this Currier and Ives print of 1861, The Life of a Fireman.

English engines preferred on the whole to produce the long, steady stroke of the slow-running engine, which they believed enabled them to regulate the flow better and to throw a steadier stream of water, whatever the variations of water flow through the inlet. One author on hydraulics added, 'We have seen some [fire engines] drawing water through long suction pipes, and the pumps worked so quickly that the water certainly had not time to pass through the hose and fill the cylinders ere the pistons began to descend', with the result that the engine could easily break down.

Indeed, during the Crystal Palace trials in 1863, it was the high-speed engines that caused most of the trouble, as Young described. One of them, 'in fetching water from a depth, ran so hot in the pumps that several men of the London Fire Engine Establishment were employed for some time in pouring cold water over the pumps to cool them, so as to enable the engine to get to work; whilst their unsteadiness and the great vibration and plunging motion when at work were the common remark of all present'.

However, in the manufacture of the quick engine, the Americans were clearly far superior to the English and European manufacturers, 'inasmuch as,

from long practice and a very extended experience under various circumstances, they have arrived at a proportion of parts which is found to give the best results in practice; whilst the general design of their engines is that best suited to those circumstances under which they are employed'.

Young was also strongly in favour of the horizontal pump in contrast to the vertical pump. Manual pumps had usually had vertical pumps but the first steam engines had horizontal ones. Young argued against the vertical position for steamers, in that 'it has been urged that the sand and grit get lodged between the packing of the pump piston and the barrel of the pump, in consequence of the water being both above and below it; and it is known that in practice such does occur, and that engines have been stopped thereby. Another objection is, that when run hard or up to its speed, the engine becomes very unsteady, shaking and jerking about in a most unsightly manner, so much so as to become very inconvenient in working, and more liable to a breakdown in consequence. The author has seen engines on this plan become so lively when working as to cause all the bystanders to withdraw to a respectful distance for fear of an accident.'

In America, such meticulous distinctions and considerations were equally observed, and there were plenty of trials for those who were increasingly reluctant to participate in European trials. Com-

Below: The mechanical ingenuity and fascination are only too apparent in this 1897 American steamer.

Above: An early Amoskeag horse-drawn steamer, with accompanying firemen at the ready.
Far right: An advertising sheet of about 1868 for the famous range of Amoskeag fire apparatus.

Below: Horse-drawn steamer, hose-reel and hook-and-ladder rushing to a New York fire, circa 1866.

petition and rivalry between the two sides of the Atlantic became so fierce that many of the American manufacturers finally refused to provide information to their colleagues in Britain and Europe and pressed on with their own improvements and developments. In his efforts to extract information for his book from American manufacturers, Young was greatly frustrated: 'So far as the author can gather from Americans resident in London,' he complained, 'the disgust and annoyance felt by them from the treatment they received on the visit of the American firemen and engines in 1863, and the results of the so-called "trials", have raised such a feeling amongst them as cannot fail to lead them to refuse any information, or to render any assistance on such a subject, when sought to be obtained from this side of the Atlantic.'

Towards the end of 1859, at steam trials held in Philadelphia, there was a primitive yet effective test applied: 'A mast, 173 feet in height from the ground, was erected, to the top of which a running halyard was attached for the purpose of raising and lowering a line of small tin cups fastened at a distance of one foot apart. This line of cups was drawn down when each engine had finished the trial for height, and water being found in any cup was taken as the indication of the height to which it had thrown water.' Each of the engines taking part had to work for 20 minutes. Among other engines, there were the Mechanic and Hibernia and Good Intent, built by Reaney and Neafie of Philadelphia; the Weccacoe, built by Merrick and Sons, also of Philadelphia; the Baltimore, built by Poole and

Hunt, also their Washington; the Independence, built by Hunsworth, Eakins and Co; and the South-wark, built by Lee and Larnard of New York.

Steam was raised in varying times, between 11 minutes (the Baltimore) and 18 minutes (the Washington); horizontal distances varied between 254 feet (the Hibernia) and 109 feet (the Weccacoe); vertical distances were between 181 feet (the Hibernia) and 83 feet (the Weccacoe). Water was found in all the cups when the Hibernia and the Washington made their trial, and in all the cups but the uppermost one when the Mechanic made her trial.

An interesting comparison can be made with the manual engines that had their own trial at the conclusion of that held by the steamers. Horizontal distances were obtained of between 150 feet and 196 feet when fully manned and working for only two minutes. It is very hard to believe that they would have maintained that rate for any longer, and certainly not for the 20 minutes that was compulsory in the steam trial.

The following year, Messrs Ettenger and Edmond of Richmond, Virginia, built a steam engine for St Petersburg; the export market was by no means exclusive to British manufacturers. Weighing 5000lbs, this engine was capable in trial of reaching full working within ten minutes of fire-lighting; it then achieved a one-and-a-half-inch stream jetted 250 feet. In another trial, there was proof that sheer horsepower would not always win the day: in open competition, a five-horsepower Neafie reached 260 feet, while a seven-horsepower Lee and Larnard threw only 248 feet. Also in 1860, the Fire King, built for the fire department of the town of Manchester, New Hampshire, managed to throw a one and a quarter inch stream 292 feet, an amazing distance for that time – so much so that it was recorded by Young with something approaching disbelief, but 'this was stated to be a fact by several parties who were present'.

We can gain an interesting insight into the growth of steam power in America by taking a close look at one fire department and following their acquisition of steamers. The Washington DC Fire Department had its first steam fire engine sent to the city by Poole and Hunt of Baltimore in 1858, in the expectation that the fire department would buy the engine with more orders to follow. Clearly, the fire department were not impressed: they returned the engine to Poole and Hunt after a few months. The following year another steamer arrived, this time

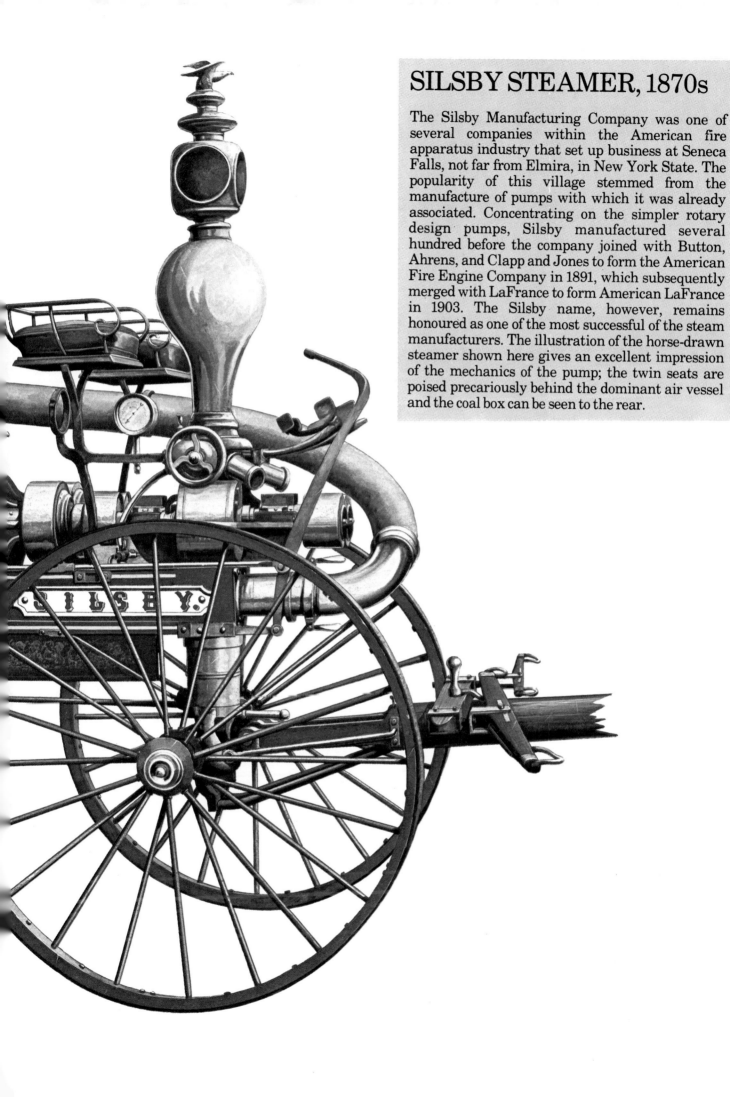

SILSBY STEAMER, 1870s

The Silsby Manufacturing Company was one of several companies within the American fire apparatus industry that set up business at Seneca Falls, not far from Elmira, in New York State. The popularity of this village stemmed from the manufacture of pumps with which it was already associated. Concentrating on the simpler rotary design pumps, Silsby manufactured several hundred before the company joined with Button, Ahrens, and Clapp and Jones to form the American Fire Engine Company in 1891, which subsequently merged with LaFrance to form American LaFrance in 1903. The Silsby name, however, remains honoured as one of the most successful of the steam manufacturers. The illustration of the horse-drawn steamer shown here gives an excellent impression of the mechanics of the pump; the twin seats are poised precariously behind the dominant air vessel and the coal box can be seen to the rear.

from the American Fire Engine Company, and in open competition with the resident Franklin manual it proved itself vastly superior.

With the outbreak of civil war, the Government stationed a Hibernia engine in the city to protect Government property threatened by the *Merrimack*, the notorious ironclad vessel that achieved undying fame for its stalemate duel with the *Monitor*. The greatest disaster that struck Washington during the war years was at the Arsenal in 1864 and was caused by an accident. Three metallic pans containing chemical preparations necessary for manufacture of fireworks for the celebrations on the Fourth of July, and which had been placed close to the main laboratory where 108 girls were preparing cartridges, were suddenly ignited by the heat of the sun's rays. There was an explosion that was so great that it threw some burning material through an open window of the laboratory, which immediately caused a fire within: 17 girls were actually burnt to death and 20 or 30 were badly injured and crippled. The fire fighters did, however, manage to save a nearby magazine from exploding as well.

Washington acquired a paid fire department in the same year. A chief engineer was appointed as well as a foreman of each company in the department and an engineman, as well as hostlers who should 'have the horses at all times ready for immediate service'. Firemen and extra men were to attend at fires when necessary. A year later, at a fire that caused great destruction at the Smithsonian Institute, and which was caused by the mistaken insertion of the pipe from a stove into the ventilating flue instead of the chimney, most of Washington turned out to watch their new fire department at work. They had also the rare opportunity of watching one thief stealing out surreptitiously from the taxidermist's room with a stuffed bird tucked under his arm: when he realized that everyone had seen his exit, he tore the head from off the bird, stuffed it into his pocket, threw the body of the bird down on the ground, and ran for his life!

By 1879, there were six engine companies. Engine Company No. 6 was formed with the addition of an 1876 Clapp and Jones steamer with a capacity of 550gpm. Four years later, Engine Company No. 5 also received a Clapp and Jones steamer but theirs had only a capacity of 450gpm. A similar Clapp and Jones was purchased for Engine Company No. 7, which was created in 1885. It was shortly after this, in the year that the Washington Monument was dedicated and President Cleveland was inaugurated, with thousands of visitors crowding the capital, that a severe fire occurred at the National Theatre. It was so completely destroyed, in the middle of the night, that the Washington Post claimed that the red glare enabled a newspaper to be read three-quarters of a mile away at the United States Capitol – perhaps that was something of an exaggeration, but it made effective headlines!

The department was soon buying Amoskeag fire engines. In 1886, an Amoskeag second-size steamer of 700gpm was received by the District of Columbia Fire Department. In the same year, the perils of the job were clearly depicted in a fire in a tin and hardware shop on Pennsylvania Avenue which

Below: The San Franscisco Fire Department has a long pedigree; this steamer was part of the department's equipment in 1877.
Right: LaFrance Metropolitan steamer, Fourth Size, from about 1890.

caused the explosion of several tanks of oil after the firemen had already arrived. Besides injuries from flying glass to sixteen people, including the assistant chief engineer, the horse belonging to the fire chief himself 'had his eye gouged out and a blood vessel in his leg severely cut by glass'. In 1891 the department received a 1000gpm Clapp and Jones, but 500gpm and 600gpm Clapp and Jones engines were still being supplied at the end of the century.

Horses for the fire department had to be carefully selected and there were many opinions as to which kind made the best and the toughest. One opinion in the District of Columbia Department was that the solid-coloured horses, especially the greys, made the most suitable horses for fire department work, because they generally had the most mild of temperaments and were reckoned to be the most intelligent. The department's veterinarian would check each horse carefully and put it to a test – generally to pull an empty carriage: 'If the horses were puffing at the end of the test, they were declared to be too short-winded for service.' There were always problems with ailments, particularly with horse influenza, which seemed to spread rapidly among horses of all departments. There was also the nasty disease

known as osteo porosis, which was accompanied by a swelling of the animal's head. As the historian of the District of Columbia Fire Department recorded: 'Whenever anyone around a fire station was heard boasting or engaging in heroics, it was the good natured practice for some firemen to shout, "Call the veterinarian – this man has got osteo porosis".'

The firemen were constantly trying to develop methods to speed up the time it took for the steamers to get out of their station when the alarm sounded. There were three main factors to consider: the horses, the men and the machine itself. The men were speeded down from their bunks above the station by the introduction of the sliding pole. The horses were stabled right beside the engine and could be backed into their traces in a matter of seconds, with a quick-hitch harness that dropped over their backs as soon as they were in position. As for the steamers, it was the priority to get up steam as quickly as possible, particularly if the fire was near and there was little time to do so on the way.

The answer to this problem was to maintain steam pressure in the boiler constantly, even when the engine was not in use. This was done in a variety of ways, one of which was to have a small stove located in the cellar of the engine house and attached to the

FIRE MUSEUM OF MARYLAND

MARION FIRE COMPANY №1

*Above: 1908 aerial ladder built by American
LaFrance (Fire Museum of Maryland).*

steamer by means of a breakaway connection,
which broke free as the engine was drawn out of its
engine house. At the same time, the chimney cap of
the boiler, which had been in place to keep in the
heat provided by the stove in the cellar, was also
dragged off by means of a link to the ceiling of the
engine house. With the gauge already showing a
head of steam pressure, the engineer started his
fire beneath the boiler as quickly as possible, even
as the appliance passed through the door of the
engine house, so that there was virtually no time for
the pressure to drop. It is not difficult to imagine
the urgency of the occasion, with each man about
his business, the excited horses, steam drifting
from the boiler as the chimney cap was raised, soon
to belch out with greater force as the furnace of the
engine was fuelled.

The end of the century saw a new peril. The
enthusiasm of cyclists led them in their dozens, both
men and boys, to ride to the fires to watch the fun.
At first only an incidental nuisance, they became
a serious danger when as many as 100 cyclists

74

Above: Firemen at work during the burning of
Chicago in 1871.
Below: 1888 Hayes patented 55-foot aerial ladder
built by LaFrance (Fire Museum of Maryland).

would pedal through the streets, intent on racing the steamer, if possible overtaking or leading it and getting in the way of the horses, causing the utmost confusion and not a few accidents, not only among themselves but among the horses and the steamers also. To cap it all, they would fling their bikes down once they arrived at the fire and the machines would trip firemen up and snag the pipelines and cause endless trouble. Eventually the Chief Engineer of the District of Columbia Fire Department was forced to persuade the police to stop cyclists from interfering with the movement of the appliance and from parking anywhere near the fire. It is doubtful if this order had very much effect: there was little time in the emergency of a fire to pay heed to those who were disobeying the order. Spectators at a fire have always had an irresistable urge to get in the way as much as possible in their overwhelming and insatiable curiosity at the proximity of such disaster!

By 1900, the District of Columbia had a total of 14 engine companies, four truck (or ladder) companies and two chemical engines; a fifth truck company was placed in service toward the end of the year, equipped with a 1900 American LaFrance Hayes aerial ladder of 65 feet. By that time there had been many innovations in design and power, both in America and in Europe.

In other parts of America, in the years before the

Above: Steamers in action at the costly Barnum's fire in New York, 1865.
Right: Amoskeag self-propelled steamer with differential gear to assist cornering.

Below: Yet another appliance in the famous line of Fire Kings; this one dates from 1905.

Civil War, many new manufacturers appeared. Besides Reaney and Neafie, there were several manufacturers of hand pumps who tried quickly to convert to steam power. Some were more successful than others. Agnew and Hunneman both built a limited number of steamers; Jefferson of Rhode Island produced more than 50 steam engines before giving up the competition; the Button Company sold more than 200 steamers; Silsby Steamers sold well, and so did those of the Amoskeag Company of Manchester, New Hampshire; there were also steamers by Ahrens, by Clapp and Jones and by LaFrance. In time, the competition was narrowed to three giants. These were the American Fire Engine Company, which consisted of a merger of several of the other companies; the Amoskeag Company; and LaFrance. Later, American and LaFrance combined to form the famous American LaFrance Company, which produced the popular Metropolitan steamer. During the Civil War, these engines proved themselves more than adequately; after the Civil War, steamers became a common sight.

It was at the notorious Barnum's fire of July 1865 that the steamers finally showed how superior they were to the manuals. There were midgets, giants, albinos, wild animals, fat ladies – all the grotesqueries of a circus famous throughout the world and much loved by its contemporaries, But on that day the greatest attraction of all was the fire itself, not least because it provided a good chance to get a

look at the new steamers in action. There were three Amoskeags and two Silsbys at the fire, as well as double-decker manuals, squirrel tails and a Shanghai.

Volunteers rushed inside, killing many of the animals mercifully. A fireman killed a threatening tiger with a blow from his axe. Half a million dollars' worth of equipment and animals was lost, but it was reckoned that the steamers had saved another quarter of a million dollars' worth. As a result of that performance, New York, where the fire occurred, had a paid fire department within a month; this was constituted with more than 80 steamers, more than 50 hose companies and eleven hook and ladder companies, as well as 500 fire fighters. This figure might be compared with the 4000 volunteers who had been required for the manuals, men who, despite their fights, were tributed as 'honest, intelligent, sober and industrious'.

Americans still strove to improve their equipment and in the process they hit on still faster methods of getting their engines off to a start. Besides the under-floor stove and the automatically lifting boiler lid, some engine houses even had burning gas jets fixed in the floor just ahead of the engine, so that when it went over the jet, which was kept permanently alight, the kindling in the furnace grate was lighted automatically. Extra coal for a long fire was carried in tenders drawn behind the steamer, or might be brought along as required in canvas bags carried by coal wagons.

Such precautions, however, could not head off the problems. In 1872, there was a plague of a contagious horse disease, a form of distemper known as epizootic disease. This caused as many as a third of the horses in some departments to be out of use, with the result that the citizens themselves had to go back to hauling the engines, but with slightly less relish than in former times. Boston, with at least a third of its horses out of action, had a fire in the commercial district which lasted for 16 hours and became known as the Epizootic Fire.

The year before, Chicago had suffered the worst fire in its history. The Great Chicago Fire of 1871 destroyed one-third of the city and lasted for 30 hours. More than 17,000 houses were burnt down, as well as hundreds of stores and factories. Where the firestorm raged, destruction was total. Photographs of the aftermath in Chicago's business district resemble photographs of the results of the atomic bomb in Hiroshima, or the remains of Dresden after the Allied bombing in the Second World War.

Chicago was immensely prosperous: apart from its position as a grain centre, a railroad centre and a shipping centre, it handled vast amounts of stock, housed thousands of immigrants from Europe, and possessed factories that churned out all manner of goods. The city was a major fire risk because so much of it was made of wood, not only the houses themselves but also the sidewalks. The fire record was already higher than New York's. On top of that, Chicago had suffered a dry, late summer, with winds blowing off the prairies to dry up what little rain did fall. A bad fire on 7 October had already

Right: American Fire Engine Company Combination Wagon of 1899, now at Lawrence, Massachusetts.
Bottom right: A 1908 American LaFrance aerial, with 65-foot extension, in service at New York.
Below: 1905 Hale Water Tower by American LaFrance (Fire Museum of Maryland).

exhausted the city's fire department; there were only twelve Amoskeags and three Silsbys in operation, besides a mere handful of hook and ladder trucks and hose wagons.

It is probable that the fire was started by accident by a citizen with a kerosene lamp in a barn. It did not take long to spread, travelling faster than a man could walk. Mobs of looters took the opportunity to raid alcohol stores and got in the way of the engines, which in any case were constantly driven back by the heat. One engine was crushed beneath a falling wall. Firestorms developed as hot air surging upwards was replaced by cold air rushing downwards, like a tornado, with the five-storey blocks in the office area acting like chimneys of fire that were quite unstoppable. The trolley tracks twisted into weird shapes with the heat, which was somewhere near 3000 degrees Fahrenheit at the heart of the fire.

As thousands of Chicago citizens fled across the Randolph Bridge, hundreds of buildings – the Courthouse, the Post Office, the Customs House, the Chicago Times building, the Chicago Rock Island and Pacific Railroad Depot, Cyrus McCormick's Reaper Factory – all these and many more fell before the fire. Looters even stole horses from the engines themselves, so that they could pull their own wagons; when the engines prepared to move on, the men had to pull them by hand. Spare engines were brought in from the surrounding

Above: A steamer was a proud possession for the Virginia City Fire Department – and many others.
Bottom right: In the town of Valdez, Alaska, built entirely of wood, a fire engine was vital.
Below: A Waterous steamer of 1886–1900. The functional bareness of the design is the attractive feature.

districts by rail – as many as 20 or 30 of them – until finally the fire exhausted itself. In so doing, it killed 300 people and caused something like $200 million worth of damage.

Not surprisingly, people became increasingly interested in fire protection. Further improvements included self-propelled Amoskeags, whose popularity was heightened after the dreaded epizootic disease: the fear of many that there would be a recurrence of the disease forced them, often against their better judgement, to put their faith at last in self-propelled steamers. Later Amoskeags had differential gears to help them round corners.

In England, one of the most popular machines toward the end of the nineteenth century was the Merryweather Fire King. This was, in fact, their first self-propelled steamer, produced in 1899. The Fire King was available in different models with slightly different specifications, depending on the needs of the client.

But engines were not all: there were also developments to the ancillary equipment that was such an essential part of the fire department's armoury. Ladder and hose tenders, often elaborately decorated by their proud companies, became more sophisticated. There were Skinner's ladders of 80 to 100 feet; there were Hayes aerials with their special crank raiser; there were Hoell's scaling ladders. In 1868, Babcock aerials were available, raised by vertical worm screws on either side of the turn-table base. There were Seagrave ladders, too, after the fashion of the Hayes aerials. In 1895, there were Dederick aerials and Pirsch hook and ladders. Greenleaf water towers in three sections appeared in 1876 and three years later were demonstrated to the chief fire officer of New York. Initially sceptical, the chief accepted the tower and ordered more after a fire in which a New York newspaper reported that the tower 'had a perfect sweep of the fourth and fifth floors and threw an effective stream over and on the roofs of the adjoining buildings. Everyone in the vicinity appeared perfectly astonished and admitted that it was the greatest thing they ever saw, and a valuable auxiliary to the fire service. The universal verdict was that a new and important apparatus for the extinguishment of fire had been added to the equipment of the fire department.' The first 'aerial' truck to be delivered to the District of Columbia was in 1902 – a Seagrave aerial with a ladder that could be raised to 75 feet.

As the age of steam reached its peak, city street escapes appeared on corners; New Haven Parmelee and Grinnell sprinklers were introduced in buildings; quadricycle hose reels confirmed the popularity of the bicycle; hand-drawn chemical engines heralded a new development of fire fighting. Steam was a period of great excitement, an age of machines and the development of power, but it was to last for only a short time. There was still more power yet to come to the aid of the fireman.

Motor Power

Steam-pumpers had barely come of age and the great self-propelled steamers were only just beginning to find acceptance when they were abruptly superseded by the internal combustion engine at the turn of the century. In terms of the relative slowness with which the fire pump had evolved until the arrival of steam power, the speedy replacement of steam by motor power was indeed remarkable. But in terms of our modern concept of the pace of change, steam did in fact die hard. Brigades in many countries were ordering steam engines as late as 1920, and rural engines were still fighting air-raid fires in Britain with steam pumps 20 years later still. Despite the advantages and efficiency of the petrol engine, local brigades that had only recently and at considerable expense equipped themselves with the very latest steam engine could not always afford to replace such costly items quickly.

There was a number of petrol-driven ancillary vehicles in use both in the United States and in Europe for some years before the first self-propelled petrol-motor fire engine was put into service by a public fire brigade. The credit for this important 'first' went to the Finchley fire brigade in England. Their engine was built for them by Merryweather and Sons, who had already won a considerable reputation for their achievements in the realms of

Left: The bulldog nose of the Mack tractor, familiar to firemen all over America.

steam. Earlier in the same year, 1904, Merryweather had built a motor fire engine for a private customer – the Rothschild estate in France – but their sale to a public brigade was of far greater and more widespread commercial importance.

Whereas earlier petrol engines had been used only for hose wagons, chemical appliances or in combination with horse-power or steam-power, the Finchley engine used a single petrol engine both for propulsion and for pumping. It was therefore able to maximize on the petrol engine's greatest immediate advantage over the steam engine – its ability to reach full power far more quickly. In addition, the Finchley vehicle was also equipped with a telescopic fire ladder, fire extinguishers to be used in emergency by hand, suction and delivery hoses, chemical equipment and first aid. It was in fact unusual among many of the earlier petrol vehicles that they should carry a variety of ancillary equipment; more often the vehicles were used basically as

pumps, other necessaries being carried by separate tenders.

The Chief Officer of the Finchley fire brigade was himself partly responsible for the design of the engine. Mr Sly put his own considerable experience into the construction of his brigade's new aquisition, which was powered by a 30hp, four-cylinder engine, with chain drive to the rear wheels. The vehicle was capable of a road speed of approximately 20mph and the engine drove a pump with an output of 250gpm. The pump was driven from a shaft connecting the engine to the gearbox through a sliding pinion. This pinion enabled the pump to be disconnected when the engine was in use to drive the vehicle along the road. The pump itself was a Hatfield type, produced by Merryweather, with a reciprocating pump with three barrels whose plungers were worked off a single crank by three short connection rods. There were two hose connections. The soda-acid chemical apparatus had a 60-gallon

Above: The first self-propelled petrol-motor fire engine in use with a public fire service built for the Finchley Fire Brigade by Merryweather, 1904.
Top left: An American LaFrance steamer.
Bottom left: This 750gpm Amoskeag steamer was the pride of Engine House No. 3, Lawrence, Mass.

tank; there were 180 feet of hose and the original wheeled escape was 50 feet long.

This fire engine had a hardy life and underwent several changes of style. Twin rear wheels replaced the original single solid rear wheels, to ensure better grip on the road surface; the steering column became raked instead of vertical; the flat radiator of later Merryweather engines took the place of the curved radiator of the first model; and a 50hp Aster engine was put in place of the original 30hp engine. There was very nearly an undignified, although useful, end to the engine when it became used as a pump in a gravel pit for a number of years, but it was rescued from this unsuitable fate and is now preserved in the Science Museum in London.

Great interest was shown in the engine when it was first accepted by the brigade, and a report was made on its first trial. The report is muted in tone, but it is not hard to perceive a quiet pride underlying the words: 'In a few seconds the escape was shipped and run to a dormitory window and the chemical hose was uncoiled and a powerful jet delivered. The fire pump was thrown into gear with the motor and, drawing through three lengths of suction hose from the swimming bath, sent a fine jet nearly over the tower, which is about 160 feet high.'

Although, perhaps reluctantly, impressed by the obvious abilities of the new motor engines, there were many who were sceptical of their reliability.

These people seized on any chance to question the ultimate value of the engines, and in tests between steam and motor engines they seemed to welcome accidental breakdowns among the innovators. For a few years – and it was for only a very few years – there was the same sort of rivalry between steam and motor that had existed between the old hand pumpers and the early steam engines back in the 1850s and 1860s. But the outcome was inevitable from the outset, and the steamers eventually went the same way as the hand-pumpers – respected but outdated.

In any case, it must have been hard to be wholly sceptical when any enthusiast's curiosity could not fail to have been aroused by the intriguing engines that appeared at the dawn of the motor age, several years before the Finchley Merryweather. Daimler had been responsible for a motor fire pump as early as 1888. Four years later, they sold one to the German town of Carnstatt. In 1895, the Honourable Evelyn Ellis bought a Daimler pump for his country estate at Datchet, the first motor pump of any kind to be introduced into Britain. He proudly

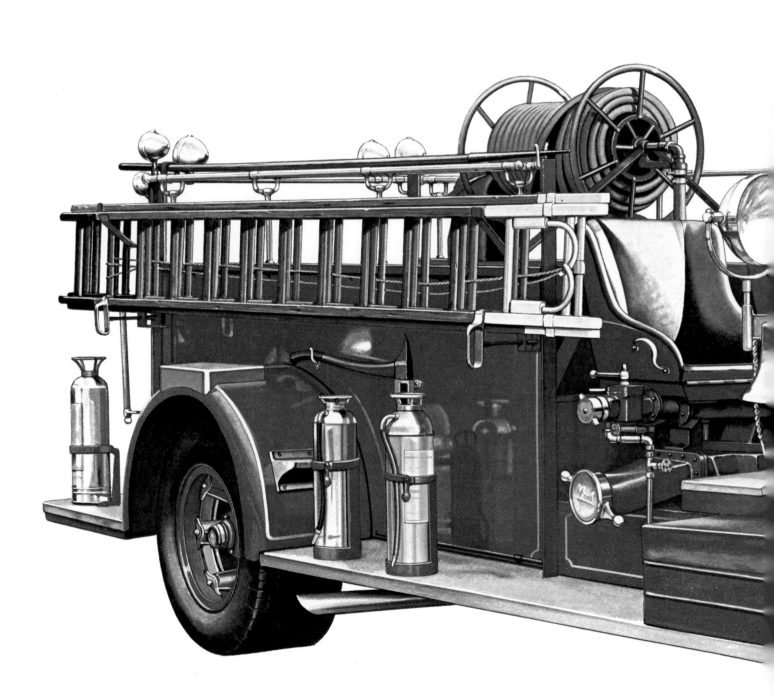

MACK PUMPER, 1935

Equipped with hose reel, ladders and hand extinguishers, this Mack pumper lacks the snub nose of the well-known AC model but the 'Bulldog' insignia is nonetheless proudly displayed over the radiator. The Mack brothers began producing chassis for trucks in 1900 in New York and later moved to Pennsylvania. One of their tractors was first used for towing fire apparatus in 1909. The AC model was introduced six years later and continued in production for more than twenty years. It gained a considerable reputation for durability with American troops in the First World War and it was from that testing period that it earned the 'Bulldog' nickname, which was adopted by the company for their insignia and has remained with them ever since. Despite the success of the popular AC model, however, Mack were sensible enough to diversify in production to include models such as the one illustrated. Their reputation for reliability was consolidated by further successes and the way was paved for such modern appliances as the Super Pumper Complex and the Mack Aerialscope.

Right: The first motor pump to be shown in Britain was this Daimler petrol engine, displayed at the Tunbridge Wells Show by the Hon. Evelyn Ellis in 1895.
Far right: A Waterous engine with inflow pipe and filter and two coiled outflow pipes.

Lower right: A 1906 pumper, well loaded up, which saw service in Wayne, Pennsylvania.

Below: The Eccles Corporation Protector Fire Engine, the first petrol-motor tender, of 1901.

displayed the pump at the first exhibltion of petrol-motor vehicles to be held in Britain, at Tunbridge Wells on 15 October.

None of these early engines was self-propelled. The Ellis-owned pump, for example, was fixed to a carriage designed to be drawn either by men or by one horse. Although, at the exhibition, it was at work within one minute of starting up from cold, there were scathing comments that its jet was no more powerful than that thrown by a manual pump worked by six men.

In the United States, too, it was horse power that was used to draw the first gasoline/kerosene-powered pumper. The Waterous pumper of 1898 was

built as an economical alternative to the very costly steam-pumpers. More sophisticated, self-propelled motor-pumpers appeared later than in Europe, although it is difficult to tie down exactly which engine can claim the honour of 'first'. In 1906, Waterous produced an engine with two motors, one for pumping and one for propulsion. Three years earlier, American LaFrance developed a motorized hose and a chemical apparatus for the Niagara Engine Company No 1, New London, Connecticut. But in 1907 Waterous came back with a single four-cylinder engine that worked both the pump and the propulsion; this engine was capable of pumping up to 600gpm.

There were other motorized appliances besides pumps. Chief Crocker of New York bought a Loco-mobile in 1901 to speed him to local fires. In the same year, the borough of Eccles, Lancashire, bought a motor tender that could carry five men at about 14mph. This was self-propelled and it was made by the Protector Lamp and Lighting Company, who also made a little car nicely called the Bijou. On its very first call, the Eccles tender was delayed by a flock of sheep and some cattle that were blocking the roadway – not unusual for that neighbourhood. Later, the tender struck an iron pillar while one member of the crew was learning to drive (driving licences were not required for motor vehicles for many years). In the same year, another tender was superimposed on a Daimler chassis and tested against horses with some success. The tender was christened Farting Annie – somewhat disrespectful but hiding, perhaps, a hint of affection.

In one test against a horse-drawn machine, Farting Annie achieved a draw, of which her crew were justly proud. But we may not be quite so sure that she achieved much speed, for, although she made such a sudden start that her crew were left momentarily behind, they quickly caught up and leapt aboard. In another test, no holds were barred, as Farting Annie resolved to

prove herself once and for all. As both machines went roaring down the hill on a test call, Farting Annie's driver recalled: 'The police tried to stop the traffic and to a certain extent were successful but a tram came just where I judged I could get through. I thought I could still do it but the back wheels caught in the tram lines and we hit the side of the tram, tearing out most of its side panel. We could not stop this time and on we went past the Adelphi, up Hardman Street level with the galloping horses, using every atom of power that we could possibly get out of our engine. At last, as we gained the top of the hill, we drew ahead faster and faster. With our tons of weight behind, it would have been all up with us if we had hit anything but we took the risk. The poor men behind were hanging on for their lives as the engine rocked from side to side. With every yard we went we were gaining ground and at last we arrived and there was the Watch Committee waiting. We had won!' In later tests Farting Annie proved herself less reliable, but she had made her point. Her driver's cry of triumph was like a clarion call to the new age.

In 1903, Tottenham borough in London built a new fire station with no provision for horses, and they bought a motor escape with a chemical engine

to put in it. This was a more fundamental change in construction and attitudes to the engine house than any of us now can probably realize. It was a reflection of changes occurring throughout Britain and many other countries, changes that deeply affected centuries-old traditions of horse power and, in many cases, brought about rapid and disconcerting changes in society as well. The age of the stable was fast disappearing. In 1911, the District of Columbia also experienced its first local engine house to be built without provision for stables or manure pit. This innovation caused a stir similar to that in Tottenham. Engine Company No 24 acquired the first piece of motorized equipment in the District in the form of a 500gpm pumper built by Waterous. Like many of the great steam-pumpers, early motor-pumpers easily earned themselves nicknames, and Engine Company No 24 was no laggard in finding one for their Waterous. Big Liz became her name, but only six months later Big Liz was replaced by a 700gpm Ahrens Fox pumper. Realizing just how useful the new motorized pumpers could be, brigades and companies all over the world were keen to try out the very latest in a fast developing field.

There were many experiments and many different combinations and adaptations of equipment in the years before the First World War. Old horse-drawn steam-pumpers were given petrol-driven motors or even battery-electric motors for propulsion. As late as 1920 there was a Teudloff Dietrich horse-drawn water tender, with a petrol engine for the pump, in Hungary. Much of the fascination of this period is the imaginative application of new sources of power before manufacturers settled down to the proved and most efficient methods. Among the earlier machines it is possible to look at only a few, but these few suffice to demonstrate the wealth of variety and the international spread of the new age.

Many of the names still famous today in fire fighting were already in evidence at the beginning of the century. Typical of the combination vehicle

Above: Teudloff-Dittrich tender, Budapest, 1918.
Left: 1899 American 1000gpm steamer linked to a
1912 Christie tractor in a neat conversion job,

being produced by these firms was an American LaFrance which consisted of a two-wheeled tractor bolted directly to the steam-pumper of an old horse-drawn pump with the fore-end of the engine removed, including the driver's seat and the controls. The result was a rigid hybrid of steam power and 105hp, six-cylinder petrol power. American LaFrance produced similar tractors of varying power for different duties. For example, they built a 75hp tractor to tow some of their aerial ladders. Several of these combinations had such a long wheelbase – this could be more than 30 feet long – that they required a steersman for the rear wheels. He sat perched above the ladders, at the back of the truck.

Besides American LaFrance, the United States boasted names such as Seagrave, Pirsch, Ahrens-Fox, Christie and Mack. John Christie emulated American LaFrance and built similar up-front tractors to replace horses and draw old steamers. The city of Pittsburgh even combined a Christie tractor with an American LaFrance steam pump. The Christie had a 90hp gasoline engine and could pull loads up to 18 tons. The engine itself was mounted in front of the front-drive wheels, a notable feature of many Christie engines, which always looked solid and square. Christie himself had ambitions to be a racing driver but became famous for the somewhat less racy but much more worthwhile creation of fire engines. In seven years he made nearly 600 tractors, which were put into use to draw every manner of appliance.

Just as Christie's engines were distinguished by their truncated fronts, other manufacturers had their trademarks which made instant recognition easy. Most notable of these was Ahrens-Fox, with their pump set well ahead of the radiator of the vehicle. An additional water-cooler was added because the radiator was insufficient for its task during pumping. The distinctive style of the engine, with its redesigned spherical air vessel gleaming

in the reflected glare of the fire, gained for Ahrens-Fox a reputation to which the efficiency of their machines gave substance.

John Ahrens himself began the company in 1908 to produce steam pumps but, when joined by Charles Fox, decided to change as quickly as possible to the newly popular motor engines. They produced their first in 1911. Their model A was a six-cylinder engine with a two-cylinder piston pump, which produced 750gpm at 120lbs psi. By the end of the 1920s, Ahrens-Fox were producing a 1000gpm pumper, as well as aerial ladders over 85 feet.

Another characteristic front was that of the Mack tractors, produced by the five Mack brothers, who set up business in Brooklyn in 1900. Originally manufacturers of truck and bus chassis, they sold their first tractor for towing fire equipment in 1909. This was to draw a 75-foot aerial ladder. Within a couple of years they were selling their own equipment as well as the tractors to pull it. The mark of the Mack is a snub nose with a circular emblem and the letter 'M' within the circle. There were critics who complained of the ugliness of the snub nose, but they were soon hushed by the reliability of the tractors themselves. American forces used the tractors in Europe during the First World War and favoured them with the name of Mack Bulldogs, a name that the company was justly proud to bear.

Seagrave came up with a motorized fire engine on the Pacific coast in 1909. This was a chemical rig, Chemical Engine No 5, which was in service for 21 years at Pasadena, California, equipped with hose, chemical apparatus and ladders, and with the two-man front seat ahead of the motor. Seagrave hook and ladders, with 75-foot extensions, were also motorized and in use at that time. The company itself was founded in 1907 and soon experimented with a front-wheel-drive electric engine as well as later producing more orthodox six-cylinder engines with mid-mounted pumps.

The Seagrave company was also responsible for major developments in the use of the centrifugal pump for fire fighting. This was a revolution in fire-fighting design and quickly became commonly used. Reciprocating pumps of the kind used in manual fire engines had continued in use throughout the age of steam, even though the first rotary pumps were introduced as early as 1785 and the centrifugal pump was produced in 1851. But steam engines did not produce fast enough revolutions to take full advantage of the centrifugal pumps, which had to wait for further development through the internal combustion engine. Piston pumps were still used by early Ahrens-Fox machines and rotary gear pumps were used by American LaFrance, among other manufacturers. In time, virtually all manufacturers turned to the centrifugal pump.

The principles of the rotary and the centrifugal pumps are simple, and yet they were in effect the first new types of pump to be introduced since Ctesibus's air-vessel pump, 2000 years earlier. Centrifugal pumps are a category of pump entirely on their own. Rotary pumps, on the other hand, are basically of the 'positive displacement' type of pump, a type that generally includes the old force pumps and lift pumps used in ancient manuals as well as 'bucket and plunger' pumps, which are a combination of both force and lift pumps.

The bucket and plunger pump is itself interesting. Like the lift pump, it consists of a cylinder with a hollow plunger ; at the bottom of the cylinder is an inlet valve. Water brought in through the inlet

valve and retained within the hollow plunger by another valve is lifted within the plunger when the plunger is raised. Some of this water passes out through the outlet at the top of the cylinder when the plunger reaches the top of its stroke. But some of the water also remains within the cylinder. This water is pushed out through the outlet by displacement as the thick trunk, fixed above the plunger, descends within the cylinder on the downstroke of the plunger.

Such pumps are also known as 'reciprocating' pumps because of the 'to and fro' or 'give and take' nature of their action. A sophisticated form of reciprocating pump is contained in a hexagonal casing, within which is a crankcase. Connecting rods working from a single crank operate plungers in three working chambers spaced equally about the circumference of the hexagonal chamber. The chambers are joined by water passages within the wall of the crankcase. By means of valves and the reciprocating pulsations of these plungers, the water is drawn from the suction passage to be pumped into the delivery passage. Air vessels are also fitted on the suction as well as the delivery side of the pump to overcome the inherent pulsations of the water flow.

The advantage of the rotary pump is that its action produces a continuous flow of water without any troublesome pulsations and without the need for air vessels, as in other types of positive displacement pumps. The rotary gear pump is, however, only of real use where relatively small amounts of water have to be pumped. Within the pump, two wheels with interlocking teeth revolve in opposite directions within a mutual casing. Water from the inlet is drawn up toward the wheels and passed around their perimeter by the turning of the teeth, whence the water is thrust through the outlet. This kind of pump is susceptible to wear between the gears and the casing; any severe wear quickly allows an unacceptable degree of leakage past the teeth, which reduces the pressure. Water polluted with abrasive elements is particularly destructive to the gears.

The centrifugal pump is quite different. Instead of such easily damaged parts as pistons, valves and plungers, it uses centrifugal force to expel the water. In simple terms, water drawn in at the centre of the pump is thrown out with great force at the periphery of the pump. It is, of course, the same force that keeps a carnival-goer pinned to the side of a rapidly revolving 'wall of death'. The water in the centrifugal pump is directed by a number of radial vanes fixed in circular side plates. This device is known as an impeller. The speed of the water increases as it is flung out toward the rim of the impeller. The partial vacuum created by the flow of the water out from the impeller draws in more water through the central inlet. The water leaves the impeller in the form of kinetic energy, or velocity; this is then transferred within the casing to a different form of energy – pressure energy – providing the powerful, steady flow that the fireman needs.

These were the types of pump being fitted to the engines of the new motor age, both in America and in Europe. But standardization of any kind took a

long time. Europe had as many varieties of engines and vehicles as the United States. In Europe, there was, for example, the 1907 Delahaye-Farcot, well-loaded with hose reels slung to the sides and pumping equipment, capable of carrying 15 men into the bargain. Popular in France, this useful appliance was sold abroad as well. Delahaye were one of the very earliest French motor car manufacturers but they readily adapted themselves to the requirements of the *Corps de Sapeurs-Pompiers* in many French towns, as well as to the needs of the capital. Among the range of appliances built for Delahaye for use in Paris was their 1926 rear-mounted pumper; this also had large water tanks and detachable hose reels, and a capacity of 250-400gpm.

From Italy came a fire pump based on a 35hp Isotta-Fraschini chassis, built in Milan for the Turin fire brigade, with the usual rear-mounted pump and crew seats facing outward according to the Braidwood body style of the previous century.

This body style remained popular, although it became increasingly dangerous as vehicles became ever faster; it was quite common for firemen to be thrown out of their seats when rounding corners or when avoiding obstacles, however hard they tried to cling on.

Daimler coupled with Porsche in 1914 to produce a 60hp petrol-electric pumper with a pump mounted in the centre of the appliance and a hose reel that could be demounted. This was, apparently, a more popular aspect of appliance equipment in European countries than in Britain. Merryweather, for example, supplied a rear-mounted pumper with demountable hose reels to the Cordoba fire brigade – something they would possibly not have been called upon to produce for their home market.

Daimler-Benz equipment of the 1920s was popular in its own market. A typical appliance, supplied to the Stuttgart fire brigade in 1928, was powered by a 68bhp engine which drove a 440gpm pump. It was,

Above: The pumping mechanism of a 1938 Morris Merryweather.
Far left: 1913 Seagrave with hose and ladder.
Left: 1933 Ahrens-Fox 1000gpm pump, with six-cylinder motor and characteristic front.

in fact, electric power that took the fancy of many German manufacturers in the early days, and this form of power was exploited more in Germany than in any other European country. A good example of German ingenuity was the 1910 'elektromobil-steam' fire engine built by Braun of Nuremberg. The steam provided the power for the pump; electricity provided the propulsion for the vehicle. The electric motors were built into the front wheels. The pump itself was a two-cylinder reciprocating pump which produced 330gpm and was mounted in front of the boiler. The boiler used paraffin fuel; the flow of the fuel was made smoother by the pressure of gas produced from carbon-acid cylinders.

Carbon-acid was used for other purposes by Braun. In an electric-powered vehicle supplied to the fire brigade of Charlottenburg, cylinders of carbon-acid were used to discharge the 110 gallons of water carried in two great tubes of metal fixed along the entire length of the vehicle beneath the twelve-man

Above: 1936 Hestair Dennis, still with the open bodywork. Dennis rapidly became one of the dominant fire engine manufacturers in Britain with a wide range of appliances. They are still leaders in fire-engine design.

crew. Two Siemens-Schuckert 7.5hp motors provided the power. In the Braun-Schappler extending ladder, traction batteries were used to power the vehicle and to turn and elevate the ladder itself, but carbon-acid was used to provide the pressure to extend the ladder.

For several years it was common practice to combine electric motors fixed within the front wheels for propulsion with a steam or petrol pump. One such electric-petrol engine, made by Braun again, was built for the Hanover fire brigade in 1911. The prominent feature of this appliance was the 54hp, four-cylinder Argus petrol engine mounted at the rear, with the radiator facing toward the rear; the pump outlet itself was in the middle of the appliance. The pump was a Pittler-type and could deliver 220gpm. With the radiator at the 'wrong end' of the appliance, a modern enthusiast might be forgiven for a certain degree of confusion as to which way the vehicle was facing!

As with Delahaye in France, Laurin and Klement in Czechoslovakia also turned their skills in car manufacture to the construction of fire engines. The firm began at the end of the nineteenth century, producing cycles and motorcycles. In 1913 they produced a fire appliance with a four-ton chassis powered by a 32hp four-cylinder engine. The positive displacement rear-mounted pump produced 300gpm. Apart from the extension ladder slung overhead and the coils of hose also hanging from the same overhead points, the most remarkable feature of the appliance was the Braidwood-style body with unusual angled seats.

British vehicles of the first two decades of the century included an equal breadth of inventiveness. Among the famous early names, Dennis delivered their first motor appliance in 1908 to Bradford in Yorkshire. Quick to take advantage of new techniques, Dennis equipped this appliance with a single-stage centrifugal pump bought in London. Later Dennis designs incorporated a turbine pump originally designed by an Italian engineer. The output of Dennis designs increased very fast. In that year alone, seven more fire engines were built; two years later, the number was 27; twenty years later, 129 appliances were built, many of which were sent abroad.

Merryweather, too, tried out the new turbine machines. Their first demonstration vehicle made use of a German Tetzel turbine pump which could be interchanged on the vehicle with the popular Hatfield-type pump – a name familiar from the first Finchley brigade motor pump of 1904. Also popular were front-mounted pumps which could be driven direct from the road-engine crankshaft. Romford fire brigade in Britain had a British-assembled Fiat F2 of this type in 1922; Southampton fire brigade had a similar front-mounted pump on their Oldsmobile. In both vehicles, fire-fighting equipment was supplied by Dennis.

Dennis even equipped Rolls-Royce conversions, once again with a front-mounted pump affixed before the dignified radiator. Such an appliance was supplied to Hong Kong in the decade before the

Above: From Austria came this idiosyncratic Feuerwehr Elektromobil Gasspritze of 1903.

Below: More than just a first-aid vehicle, this appliance was equipped with breathing apparatus and a pump.

Second World War. Other cars were also adapted to fire brigade use. A Morris Commercial six-wheeler appeared in 1931; Triumph cars produced a chemical appliance in 1932; in 1927 Delage produced a pump and CO_2 trailer; Ford did much the same five years later.

In the interests of safety for the crewmen, the Braidwood body gave way to the New World type of body in which the crew sat facing inward – and no doubt felt a great deal more confident. This style could be seen, for example, in the Dennis FS6 bought by the Birmingham fire brigade in the 1930s. The appliance sported a 900gpm pump which was mid-mounted, together with twin first-aid hose reels. The New World body could hold 12 men. At the same period, the Edinburgh fire brigade obtained an appliance with an enclosed body. This was possibly the first of its kind, although others might claim this privilege for the later Dennis provided for Darlington fire brigade in 1931. In the latter case, there was a rear-mounted pump with considerable ancillary equipment. Not surprisingly, the enclosed appliance rapidly became popular, although by no means universal. There were many

clear advantages for the crew. Subjected to the vagaries and inclemencies of the weather and the perils of travelling at speed in an open vehicle, they had often arrived at fires in poor shape to make their utmost effort. There was no question that the enclosed appliance would make them soft; with what they had to tackle, that was never likely. But at least it gave them a chance to prepare themselves more fully for what lay ahead. That, after all, was primarily their job: to pit themselves against the fire, not the journey to the fire!

Even Dennis, however, continued with open appliances for a considerable time and, indeed, two of their most popular and versatile machines of the 1930s were open. These were the Big 4 and the Big 6, which found favour abroad as well as in Britain. The Big 4 had a 90bhp, four-cylinder engine and a rear-mounted, two-stage turbine pump which was capable of producing 800gpm at 70lbs psi. There were two transverse benches for the crew and a wheeled escape. The overall look was of solidity and reliability: there were no frills on the Big 4. Nor were there any on its brother, the Big 6, which had a 115bhp, six-cylinder engine, also with a two-stage turbine pump, which could be fitted either at the rear or amidships, depending on what each particular customer wanted. Rear-mounted pumps in association with wheeled escapes could prove an impossible combination at times, when it was necessary to get at the pump immediately on arrival at the fire and there was the possibility of being thwarted by an over-slow slipping of the escape.

Across the Atlantic, meanwhile, motorization had gone apace. Before the end of 1915, New York was 50 percent motorized and even the Model T Ford had been dragged into service as an adapted fire-fighting vehicle. One such adaptation, made in 1915, turned the Model T off the roads and onto rails as a first-aid spotter vehicle, with a brief to tour up and down the railroad carrying the minimum of equipment and prepared to raise the hue and cry for heavier assistance in the event of sighting a fire. Dennis and Ford made a surprising combination in the 1921 Baico Super Tonner, in which a Dennis No 1 pump was front-mounted on a Ford Model T long-wheeled chassis. But progress was not always a foregone conclusion. Only three years before New York achieved its 50 percent motorization, the city had ordered 28 new steamers. It was the last big order in America and it seemed a shame that those powerful giants, like the dinosaurs of old, were doomed to sudden extinction in the face of bewildering evolutionary developments.

Pumps were by no means the only appliances with which experiments were being made in those early years. Hose tenders based on automobile chassis were very popular, particularly in urban areas where there was enough mains pressure to supply the hose with constant water. Hose tenders were also used to carry firemen, often as many as

12 at a time. First-aid vehicles were considered essential, equipped with such basic necessities as hand fire extinguishers, axes and short lengths of ladder. These vehicles were supposed to get to the fire as quickly as possible, with a small crew, and take what immediate action they could before the main pump arrived, or in the hope that such limited first aid might be sufficient to cope with the situation. Their equipment, on the whole, together with their crew, was as much as any converted car chassis might reasonably be expected to support and yet still maintain speed without cracking up. It was a wonder that some of them did not, in fact, collapse! More and more equipment was trustingly loaded on to the first-aid vehicles. Extra hose, tools of all kinds,

greater lengths of ladder, chemical extinguishers, hand extinguishers, lights – they were piled on wherever a space could be found on the groaning chassis. Only in the 1920s, and certainly by the 1930s, did the popularity of the first-aid vehicle decline in favour of the combination appliance, which brought the ancillary equipment as well as the main pump simultaneously to the fire with the utmost possible speed.

Examples of first-aid vehicles in the first two

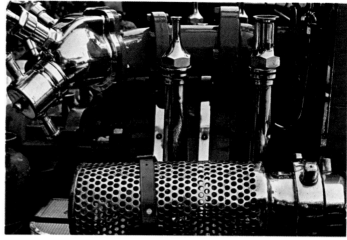

Below: The versatile and highly effective Dennis Big 4 of 1938.
Right inset: Close-up of the pump from 1914 Dennis appliance.

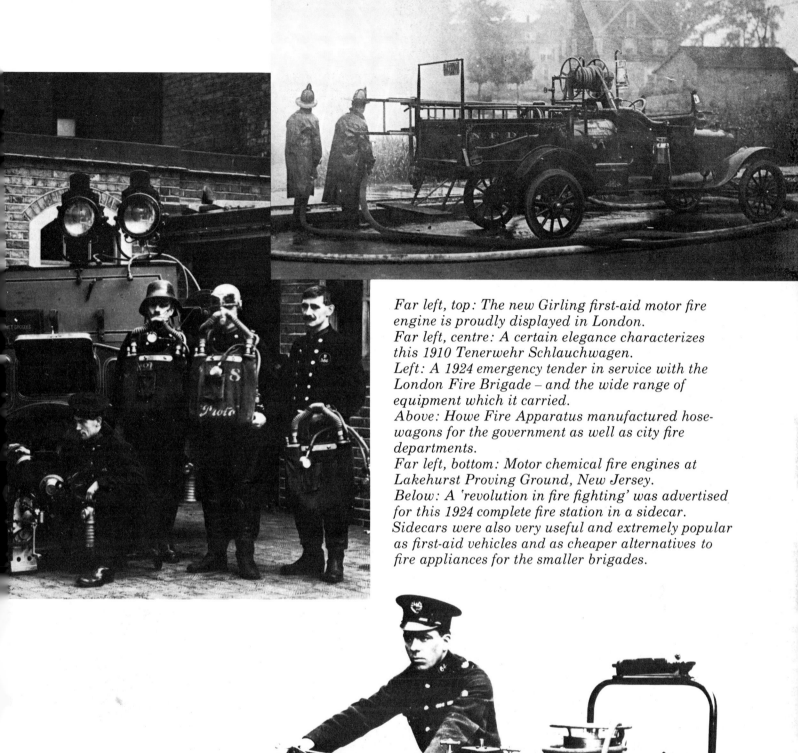

Far left, top: The new Girling first-aid motor fire engine is proudly displayed in London.
Far left, centre: A certain elegance characterizes this 1910 Tenerwehr Schlauchwagen.
Left: A 1924 emergency tender in service with the London Fire Brigade – and the wide range of equipment which it carried.
Above: Howe Fire Apparatus manufactured hose-wagons for the government as well as city fire departments.
Far left, bottom: Motor chemical fire engines at Lakehurst Proving Ground, New Jersey.
Below: A 'revolution in fire fighting' was advertised for this 1924 complete fire station in a sidecar. Sidecars were also very useful and extremely popular as first-aid vehicles and as cheaper alternatives to fire appliances for the smaller brigades.

decades of the century included a 1908 Packard with a modified body and suspension enabling it to carry as many as 15 men, together with light equipment and hand extinguishers. This model was owned by the Detroit Fire Department. In Australia, the 16hp model A3 Albion was pressed into service with the Milton Volunteer Brigade of Brisbane; it became a hose tender, equipped with a short ladder and weighed down by eight crewmen. In the Netherlands, the Model T made another star appearance: the one-ton TT truck was used in 1917 as a crew and equipment tender with a trailer pump.

Motorcycle combinations were another variation of the new theme that quickly became extremely popular. In South America, there appeared in 1913 a Merryweather conversion equipped with a hand extinguisher, an extension ladder, standpipes and a pump driven from the engine of the motorcycle itself. Leyland produced an advanced design with a separate pump engine, together with suction and delivery hose and a rather dainty transverse seat with a back-rest for a crew member. This machine was based on a BSA vee-twin machine. Later, Dennis provided appliances for a Matchless machine. There were many, many other combinations.

Chemical appliances, too, took to the petrol engine. The chemical appliance was very much a 'first aid' appliance, which in steam form had become extremely popular towards the end of the nine-teenth century. The basic form remained very much the same in the petrol-driven chemical appliance: for example, a 60-gallon drum of water that could be used on the spot without the need for a pump. The water was expelled by carbon dioxide gas created by the action of sulphuric acid with a bicarbonate solution; the result was a reasonably powerful jet, which was an effective front line of attack until the heavier jets of the main pumps could be brought to bear on the fire. Separate chemical engines were still in use in the 1930s.

There was one considerable disadvantage to these appliances: once they had been turned on, they could not be stopped until they had completely emptied themselves. If only a little was needed, the rest was wasted, and there was none left in the event of another outbreak at a separate point in the same building. There were two main solutions to this problem. Either the water tank was put into contact with several bottles of the necessary chemicals so that each bottle would expel only a limited amount of the water and the next had to be turned on before more of the water would be expelled, or twin tanks

Right: This 1912 Buick soda-acid pump would appear to be a typical automobile conversion.

Below: The streamlined appeal of the 1926 American LaFrance pumper gives credence to its power.

were used, one with water and the other with compressed air, contact between the two being regulated by a tap.

With their four-cylinder petrol engine and hydraulically operated hose reel, Merryweather claimed they could get one of their earliest chemical appliances ready for action within about five seconds of its arrival at a fire. Rival manufacturers each had their own strengths. Packard claimed a 25mph roadspeed for their commercial model with a 15hp, two-cylinder engine. In 1907, one of these models went to Ventnor, New Jersey, equipped with two 60-gallon chemical extinguishers laid transversely in the centre of the vehicle, with ladders and hose reels on top and smartly painted coachwork. Equally smartly turned out was Chemical No 1 of the Radnor City Fire Company of Wayne, Pennsylvania. In this case the two chemical extinguishers were stocked beneath the driving seat at the front of the vehicle. For those who seek, for their own sense of security, to have a solid wedge of automobile in front of the driving wheel, this 1908 vehicle would have appeared impossibly perilous: the driver sat right over the front wheels and, apart from the vertical steering column, there was virtually nothing except a lamp between himself and the other vehicles on the road.

Twin extinguishers were common. A 1919 product of the Autocar Equipment Company of Buffalo,

New York, was a 40hp chassis for 12 men, two chemical appliances and 150 feet of hose. This made the mere single 40-gallon chemical appliance made by Seagrave for Denver Fire Department in 1910 look most inadequate, but the Seagrave compensated by carrying 1500 feet of hose, an extension ladder, and an ordinary ladder, providing therefore a more useful dual role. Another single appliance was the very early carbonic-acid appliance supplied in 1903 to an Austrian brigade. This was a prime example of a single water tank with twin carbonic-acid cylinders. The water tank carried the formidable quantity of 123 gallons.

All manufacturers were keen to advertise the quality of their wares as well as quantative statistics, so when Merryweather came out with a battery-electric chemical engine with a 35-gallon chemical extinguisher and a compressed carbon dioxide gas cylinder, they displayed their pride in the achievement on the side of their appliance in bold lettering: MERRYWEATHER IMPROVED ELECTRIC FIRE ENGINE.

Improvements were, in the main, the result of experience, and no experience could rival one of the worst fires of the first decade of the century – the fire that resulted from the great San Francisco earthquake of 1906. By then, motor appliances of all kinds had already appeared, but improvements always take time to spread and in those days they took a great deal longer to do so than today. In Kensington, London, for example, in the same year, Merryweather had produced a smart new motor appliance for private estates. This was a pump that could be driven from the back wheels of a private car. Drawn on a trailer by the car, the pump was set down at the fire beside a ramp; the car was driven backwards up the ramp so that its rear wheels rested on the driving wheel of the pump; the car was then given full throttle to provide the power for the pump to produce the required jet of water. Even had such an appliance been available in San Francisco, or indeed any of the larger motor appliances then in production, it is doubtful that they could have done much better than the steamers with which San Francisco was largely equipped. And those steamers were certainly stretched to their absolute limit: Clapp & Jones, Amoskeag, American LaFrance – they were proven appliances but scarcely sufficient for such a catastrophe.

The disaster struck on Wednesday morning, 18 April 1906. There had not been earthquakes in the vicinity for some years, and it was already feared that stresses might be building up for a disastrous eruption. Baltimore and Chicago had recently experienced severe conflagrations and the fire risk in San Francisco was particularly high. As much as 90 percent of the city's buildings were wooden, crowded and set on hills – no help to horses and steamers straining up the slopes in frantic efforts to reach many parts of the city. A committee investigating the dangers of fire in San Francisco had reported that the city 'had violated all underwriting traditions and precedents by not burning up. That it had not done so is largely due to the vigilance of the fire department, which cannot be relied upon indefinitely to stave off the inevitable.'

For the 450,000 inhabitants of the city, the inevitable occurred just after five o'clock in the morning. The first shock lasted 48 seconds and was followed by four more tremors. Fire broke out almost immediately as power cables snapped and kerosene lamps overturned in dry, wooden houses. Chasms opened in the ground, swallowing whole buildings. Buildings of brick fell as easily as those of wood. The rumbling and crash of tumbling masonry and splintering wood mingled with screams of terror, cries for help and the growing roar of the flames. There was pandemonium.

Worst of all, for the fire fighters, the main water pipes broke, so that they did not have enough pressure for their pumps. Most of the hydrants failed to work. There was still the bay, from which water could be obtained, but the fire-fighting system was not geared to resort immediately to

Right: Water tower, pumps and a ladder at the tragic Triangle Shirtwaist fire, New York, 1911. Below: Combination Auto Chemical Ladder Truck built by Parkins & Co., July 1907.

AMERICAN LAFRANCE, 1917

This beautiful engine with monitor pipe has been restored to immaculate condition and now rests in the Fire Museum of Maryland. Proudly stamped on the side are the initials of the Fire Department of New York, with which it once served. Formed from a combination of companies at the turn of the century, American LaFrance spent the first decade of the new century building steamers, although the company was constructing hand pumps as well right up until 1920. The turning point came in 1910, when the company started to design its own six-cylinder gasoline motor. By 1916, American LaFrance were producing a range of appliances that included 1000 gallon gear pumpers, 900 gallon piston pumpers and 750 gallon centrifugal pumpers. In the 16 years from 1910 to 1926, American LaFrance sold and placed in service more than 4000 gear pumping engines, more than 40 piston pumping engines and five centrifugal pumping engines – a fine record of achievement.

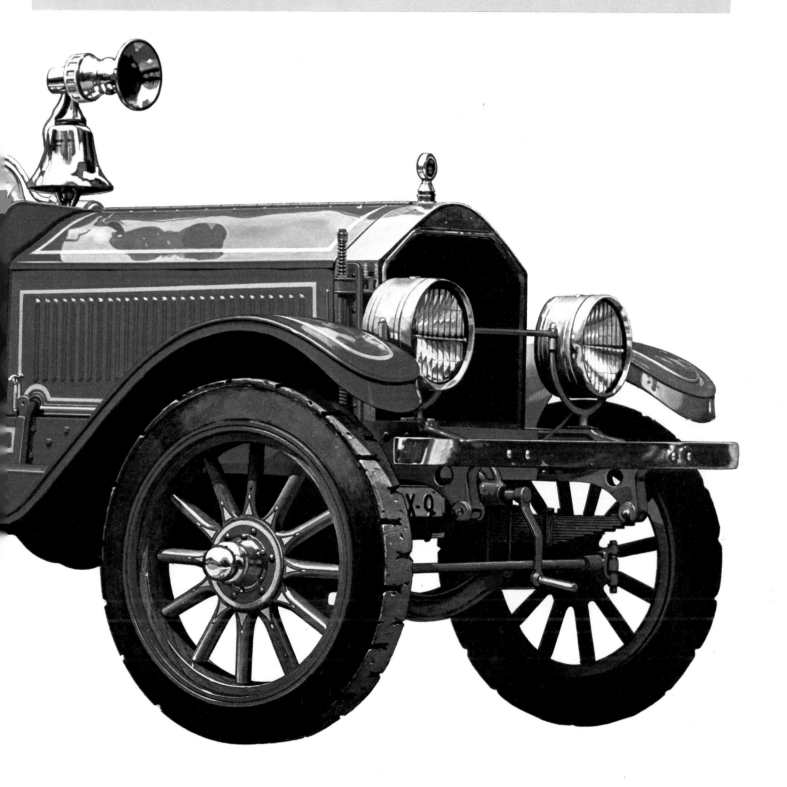

that. And then the fire chief himself was fatally wounded by collapsing buildings in the first few minutes of the disaster, leaving command to his able deputy, Dougherty. It seemed that everything was against San Francisco surviving at all.

The entire force of the fire department was mustered to attempt to hold the number of severe fires that were now spreading through the city and that quickly put the tenement area of the city beyond any hope of salvation. In all, there were 40 steamers, eight chemical engines, hose wagons to accompany the steamers, ten Hayes aerials and a Gorter water tower. Deployed about the city, they could not stop the fire jumping their first lines of defence, and the seriousness of the situation was compounded by misguided attempts at dynamiting buildings in order to clear a path in front of the fire over which it might not pass. The dynamiting only started even more fires.

The flames from the conflagration were said to have been seen more than 50 miles out to sea. The heat, the confusion, the fear of citizens stampeding through the streets, blazing Chinatown, the crash of artillery clearing a firebreak through the buildings, the desperate weariness of the fire fighters themselves struggling against what seemed to be impossible odds, the sense of calamity and despair as people watched their homes and all they possessed vanish forever in the flames – all this engulfed San

Francisco throughout that Wednesday and Wednesday night and throughout the next day as well.

Finally Dougherty and his firemen determined to make a last stand at one of the widest roads in the city – Van Ness – hoping that they could stop the flames leaping over the open space and catching a grip on the other side of the road, as yet one of the few areas undamaged by the fire. For once, luck was on their side; there were some few hydrants in working order along Van Ness. All available steamers were brought to the area and Dougherty urged on his tired fire fighters to draw on their last resources of energy in order to stop the fire. It seemed that their luck was not going to last. At one point the flames licked across the road, high overhead, and seized on a church on the other side. This fire was quickly put out with the help of a Hayes aerial, but another fire, started by another lick of flame, caught hold and was only extinguished after a hard struggle. Then, at last, it seemed that they had control of the flames. The fire they faced began to burn itself out. There was no longer any danger that it would cross Van Ness. The steamers, belching smoke and jetting streams of water into the fading flames, had at least saved some part of the city, but the fire had taken an appalling toll. It had, in the end, raged for two days, destroyed nearly 30,000 buildings, injured 3500 people and killed 674.

Firemen everywhere swore to learn their lesson that time, but it seemed that lessons were only learned the hardest way of all, by bitter personal experience. For it was only three years earlier, at Chicago's Iroquois Theatre, that almost as many people died when the supposedly 'absolutely fireproof' theatre went up in flames. Every effort had been made to provide the theatre with adequate exits and every known precaution had been taken against the possibility of fire. Smoking, for example, was banned except in a special smoking room, and there was an asbestos curtain to drop in front of the stage in the event of fire backstage. Warned by a disastrous fire in Brooklyn, nearly 30 years earlier, in which 300 people had died, the owners of the Iroquois had resolved that there was going to be no repeat of such a catastrophe. In the event, twice that number died, many of them trapped within the building by inward-opening doors, many more in a desperate attempt to leap clear of the flames from escape doors that led nowhere but straight down into a narrow alley and on to the hard cobbles below. There was no room for ladders to be raised in the alley. There was too much smoke and flame for the panic-stricken jumpers to see the safety nets held out for them by their rescuers. With the flames and their fellows pressing behind them, more than 100 people jumped to their deaths; the only survivors were those who were cushioned by the bodies of those who had fallen before.

As a result of the Iroquois fire, even more stringent rules were laid down for fire precaution in theatres throughout the world. In the United States, these last precautions paid off, at least in theatres; there has not been a serious theatre fire in the United States since that dreadful 30 December 1903. But the experience gained at the Iroquois was of little help in the San Francisco earthquake fire and of no help at all in 1910, when 146 people died in the infamous Triangle Shirtwaist fire.

Adequate space for ladders high enough to reach the panicked audience and capacious enough to convey them to safety might have lessened the Iroquois disaster. Paradoxically, it was only the year before, in 1902, that a German firm had produced a major advance in ladder construction in the form of the first turntable ladder. This could be turned through 360 degrees and could be telescoped out by compressed air to 80 feet and retracted to the length of any normal ladder. The appliance was horse-drawn, as had been one of its antecedents, a 90-foot German Magirus of 1892. Compressed carbon dioxide had also been used on another experimental machine in 1901 and Magirus produced the first turntable with elevation powered by the road engine in 1906.

European turntable ladders and American aerial ladders – each side of the Atlantic preferred its own slight variations to a common theme – were very different to the old-style wheeled escapes that later came to be associated with pumps to form the combination pump-escape that is still in frequent use. The escape van, as the separate wheeled escape was at first known, had its origin in the wheeled escape invented by Abraham Wivell and made increasingly popular with the arrival of the motor engine. Merryweather were responsible for the first motor escape van in 1903, which they produced for the Tottenham Fire Brigade, England. As well as the big-wheeled, 50-foot escape, the vehicle, or van, carried a hose reel and a 60-gallon extinguisher. Merryweather soon produced many more escapes including one, the next year, for the Bombay Fire Brigade. Many of their earlier ones already combined chemical extinguishers so that the firemen on the van could take immediate action if they should arrive at the fire before the main pump. But as soon as brigades began to acquire their own motor pumps, they began to convert them into pump-escapes, abandoning the separate escape. The 1904 Merryweather produced for the Finchley Fire Brigade, with its 60-foot escape ladder and a 60-gallon chemical extinguisher, was the world's first self-propelled motor pump-escape.

Escapes are mainly of use in Britain, while other

Far left: London Fire Brigade turntable ladder. Below: 1928 water tower in San Francisco. With a height of 30 feet, it could throw a maximum of 1500 gpm up to 300 feet.

Above: 1937 American LaFrance city service ladder truck, series 400.
Right: Pirsch 55-foot electric aerial, 1937.

countries prefer turntable ladders and long extension ladders. The outstanding feature of these
escapes is the pair of great wheels at the base. These
enable the ladder to be turned more easily when
resting on the ground. Various other features of the
ladder developed equally for reasons of facility. For
example, the extension piece secured at right angles
at the rear of the ladder allowed for greater control.
It also acted as a lever to help raise the ladder when
considerable elevation was required. With the
wheels of the ladder set so far back and acting as
the fulcrum, the main weight of the ladder was too
far forward. The right-angle extension partly
compensated for this imbalance.

Another aspect was the extension ladder on the
escape. In early escapes, this extension was known
as the fly ladder and was elevated from the main
ladder by means of a block and tackle. Any
additional height required an extra length of ladder,
fixed to the fly ladder before being extended.

Such ladders, even with their large wheels and
their extension lever, could only be pitched against
the building at an angle that would enable their
base, or heel, to rest on the ground, so that they
would remain stable. Extension ladders could,
somewhat insecurely, be set up against the building
at a slightly different angle from the top of the main
ladder, but this was not a satisfactory solution.
The only real way to solve the problem was for the
main ladder to be provided with a sliding carriage
so that the wheels could be moved along the ladder

and so that the angle at the heel could be altered at will. This improvement concluded the main modifications to the escape ladder, and it remains very much the same today.

Most escapes now consist of a three-part ladder. The levers themselves are hinged so that they can be folded along the ladder when they are not in use. This enables the vehicle to pass under the lowest of obstacles without hindrance. There are small lever wheels at the heel of the ladder, and there are chocks to place beneath the main wheels when the ladder is in position, to stop it rolling; these are attached to the ladder by chains. Guy wires and lever stays also attach the levers to the main ladder to brace the whole structure. There are, besides, small wheels at the head of the extension ladder so that it can roll easily against the face of a building when it is being elevated. One other useful piece of equipment on the escape is the plumbing gear, which enables the ladder to be kept to the vertical when not on a horizontal surface. Without this plumbing gear, the whole appliance would be in danger of toppling over sideways with the weight of the crewmen if the escape was set on a steep hill

and the ladder extended to its utmost. The purpose of the gear is to allow the main ladder to pivot on its carriage, so that the ladder itself is upright while the wheels remain at an angle on the slope. The angle of pivot is generally up to about seven degrees.

There were a variety of escapes produced during the Second World War with certain modifications, and others produced since the war which we will touch on in the next chapter. As an adjunct to the pump, the escape has proved an extremely useful piece of equipment, but the turntable ladder introduced by the Germans had its advantages. Apart from its obvious ability to turn in a full circle to face any direction, it is also free-standing; that is, it does not need the support of a wall when raised. This means that it can be used as a hose platform, standing away from the building and the flames. It can also be made a great deal longer than the ordinary extending ladder which, when too long, cannot be handled with ease. Taller buildings meant that longer ladders were much in demand. This increase in the height of the 'objective' also brought about many of the advances in fire engine design that characterized the twentieth century.

In America, these requirements were met by the aerial ladder, which at first consisted most often of an 80-foot two-section extending ladder on a horse-drawn carriage. Such ladders were mounted with their base immediately behind the driver and their head facing to the rear. They were cranked round by hand on their turntable and locked into position by a pin, being extended by a circular chain. One of the main reasons that made this appliance suitable only in the wider and straighter streets of the United States was its great length; it certainly could not have been used in Britain or France, for example. Even when closed, the entire length of the aerial ladder, horses included, could be as much as 56 feet and required an articulated carriage and a steersman for the rear wheels. Improvements to this type of ladder were made in succeeding years and the ladders were also adapted for use in Britain by shortening the overall length and by mounting the base of the ladder at the rear instead of at the front of the appliance. Shortening was achieved by breaking down the ladder into three or four sections instead of two.

One of the major considerations in the construction of turntable extending ladders was the stability of the vehicle itself when the ladder was fully extended across the chassis. In this position, with the ladder fully extended and with the weight of one or two men on the end of the ladder, all weight would be lifted off the unburdened rear wheels on the side of the appliance away from the stretch of the ladder – and the appliance would be in danger of overturning. Therefore the extent to which the ladder can be stretched at certain angles and the amount of weight which it can bear must be limited by certain safety margins. These safety margins are measured by the amount of load on the unburdened rear wheel – anything less than approximately 15cwt is considered unsafe. Scales of permissible elevations at certain angles are provided on all

Top right: Leyland-Metz Turntable Ladder, 1938. The four-section steel ladder could reach 101 feet.
Bottom right: Sampson Fire truck, 1916, which made a 350-mile run from Detroit, Michigan to Alliance, Ohio.
Below: Fire fighters of San Diego, California, took to the air with their equipment in 1917 to reach far-flung scenes even quicker.

appliances and must be rigidly adhered to. The lower the elevation of the ladder, the shorter its extension must be; only at maximum elevations can it be fully extended. The graduated scales usually make provision for one- and two-men loadings at the end of the ladder so that the fireman in charge of the elevation can read off the safety margin quickly.

Aerial ladders in many guises proliferated at the beginning of the century. Two early experimental American models were an 1888 metal aerial ladder for St Paul, Minnesota, and a later one for Seattle, Washington. Seagrave towers appeared shortly after 1900; spring-assisted aerials up to 85 feet long came in 1905; in 1912 there was a Cedes battery-electric turntable ladder; and in 1919 the Austrians produced a Steyr-Rosenbauer 46-foot ladder mounted on a truck. But for many years it was still

the hook and ladder companies of the American brigades that did the main work. Hydraulic power did not establish itself until about 1936.

Many of the early Daimler turntables were battery-electric driven. In 1909, 84-cell batteries powered a four-stage ladder which was raised by carbon dioxide. Merryweather provided the power for their 1908 turntable appliance straight from the road engine. In 1912 Benz-Gaggenau combined a turntable ladder with a 176gpm mid-mounted pump. In America, the name of Pirsch soon became closely linked with aerial ladders, and they were later instrumental in advancing the all-metal, hydraulically powered aerial ladder. But in 1913 they provided a 55-foot ladder truck to Washington State – just to prove that the old hook and ladder truck had its place. American LaFrance also provided hook-and-ladder trucks for many years after turntables and

aerials were in common use. In Germany, Metz produced 100bhp vehicles in the 1930s with a solid, clearly laid out design. One-hundred-foot aerials were achieved by Merryweather in 1933 for the Hong Kong brigade and by Pirsch in 1936 for Melrose in Massachusetts. A Leyland-Metz turntable appliance pushed a four-section ladder up to 101 feet and supplied a two-stage turbine pump to deliver 500gpm to the top of the ladder. In 1941 American LaFrance made a 125-foot ladder for the Boston Fire Department.

Meanwhile, fire fighting had taken to the water. In 1925 Los Angeles gained delivery of their pride and joy – Fireboat 2 – one of the world's powerful fireboats. Fireboat 2 was capable of pumping 13,500gpm from 13 separate nozzles. Of this output, 10,000 gallons were produced by one single giant nozzle, named appropriately Big Bertha. Fireboat 2 remained in operation as part of the Los Angeles harbour fire-protection service long after most other pumpers would have withdrawn to a respectable retirement.

Fireboats were not new. Boston had possessed a steam fireboat in 1872 and New York had one in 1875. Chicago had the Joseph Medill, built in 1908. To counteract the tendency of the water jets to thrust the fireboat away from the fire itself, the Joseph Medill had anchors which could be power-driven into the bed of the river in order to hold it in place. Other boats used jets of water fired in the opposite direction or screws to hold themselves in position.

One of the last big American fireboats to be built before the Second World War was New York's Firefighter, which remains one of the world's most impressive and effective fireboats. Firefighter was 134 feet long, 32 feet in the beam, with a diesel-electric engine. Equipped with nine nozzles at first, it could produce up to 20,000gpm. Later, it was reduced to eight nozzles when the ninth, on its elevating 55-foot tower, was removed because of its unwieldiness. When built, Firefighter cost very nearly $1 million. A year later, in 1939, on the very brink of war, Britain produced the James Braidwood, named after the famous fire fighter of the early nineteenth century. Forty-five feet long, the James Braidwood was capable of 20 knots, built for speed more than high-capacity water power; turbine fire pumps had a capacity of 750gpm each at 100lbs psi.

Water could be used to combat ordinary fires but proved ineffective against the increasing threat of oil fires, particularly at airfields. Sprayed onto an oil fire, water will only spread the flames; foam is

Right: Navy patrol boat converted for fire fighting. Below: 1936 American LaFrance. One of four 'Duplex' pumpers built for Los Angeles Fire Department at that time, with twin 1500gpm pumps.

necessary to suffocate them. And so the foam tender became another feature of the fire-fighting scene. On the earliest truck conversions, foam compound in cylinders was mixed with water to spread over the flames. Foam tenders on their own were rare; it was more common to build a combination foam tender and pump, a multi-purpose machine that could cope with any emergency. Typical of such a combination was the 1936 Austro-Daimler tandem-drive model with a foam pre-mixer and two 176gpm pumps. On the 1938 Austro-Fiat 'AF Super' enclosed appliance with foam pre-mixer and portable as well as front-mounted pump, the portable also had 176gpm output, while the mounted pump had 266gpm capacity.

Hose layers, too, became more sophisticated as requirements became more demanding. Two basic types of delivery evolved. In the 1936 Dennis, for example, the hose was hung on special bollards inside the appliance. In this model one and a half miles of hose could be delivered at 15mph. In the other type, the hose was flaked in layers laid the length of the appliance, as in some conversions of the American Ford 91T truck.

More specialist appliances appeared with the coming of the Second World War, during which fire departments reached their most inventive as they found themselves matched against great

crises. The motor engine had given them the power to experiment with and diversify their equipment. New perils had already stretched their inventiveness seemingly to its limit. Appliances existed in the 1930s whose power had never been dreamed of 50 or even 20 years before. In 1931, the American LaFrance 12-cylinder V-block 240hp engine had a pumping capacity of 1000-1500gpm; by 1937 there were double pumpers with V-12 engines producing 2000-3000gpm with four hose-carrying manifolds, each with 3000 feet of hose. But larger and fiercer fires, in ever larger and taller buildings, preying on modern, highly combustible materials challenged fire fighters all over the world to produce ever more powerful and adaptable appliances. They met this challenge with predictable ingenuity, resourcefulness and determination. The years of war and the three decades since the Second World War produced an impressive variety of appliances that amply demonstrated that ingenuity and skill.

Top right: Foam tender with hose reel, in service with the London Fire Brigade.
Right: Dennis hose layer in service with the London Fire Brigade in 1936, carrying 1½ miles of hose.
Below: This 1930s New York City fireboat, with a maximum speed of 18mph, was capable of pumping up to 16,000gpm.

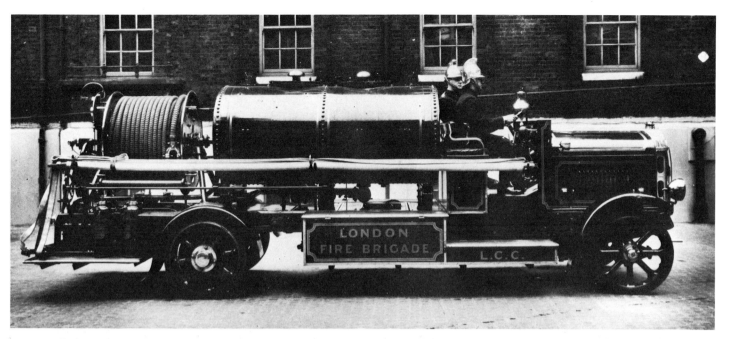

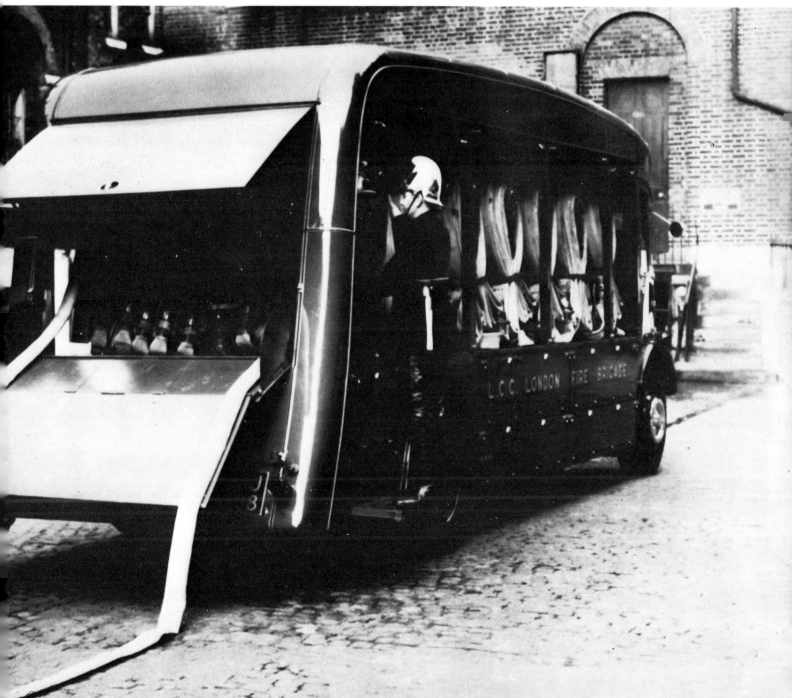

FireEngines Today

Whatever the quarrels about rearmament among the nations of Europe, with the ever-present threat of Hitler's Germany, there was open rearmament of a different kind among the fire-fighting forces of those nations, not least among the Germans themselves. As war came to Britain, the Air Raid Precaution Vehicle (ARP) and the Auxiliary Towing Vehicle (ATV) became common sights. Unable to keep up with the demand, manufacturers were rivalled by home-grown adaptations using standard truck chassis. One such adaptation was the Scammell street sprinkler, with its 1250-gallon tank, which was converted for use as a decontamination vehicle in case of gas attack. ATVs might consist of private cars or even taxicabs – the Austin Twelve saw splendid service as an auxiliary vehicle during the war, towing a trailer pump and with a ladder secured to the roof – but there were also larger vehicles, especially box-vans with room for a crew of firemen, lockers for equipment, and a pump on tow.

Mobile dams also came into popular use. Again, these were often conversions. A coach, for example, might have its back cut off, in place of which would be fixed a steel-framed mobile dam. Trucks were ideal for the job, for they could be used as dual-purpose machines for transporting wartime

Left: This 1956 Seagrave with a 150-foot Metz aerial had one of the longest reaches of all.

materials and necessities during the day and as conversions to fire service during the high-risk periods of night; for the change, all they needed to do was to take on board one of the mobile dams. These dams could hold anything up to 1000 gallons of water. The four sides of the rectangular steel frame came together to form a top that was narrower than the base; this helped contain the pressure of the water mass. Within the frame, there was a treated canvas lining. Proper water tenders usually had only about half that capacity and were very popular with the Germans.

To meet the menace of the Blitz, the Auxiliary Fire Service (AFS) was formed. By 1939, this constituted a force of about 100,000, from all kinds of work and background, all endowed with remarkable enthusiasm reminiscent of the early volunteer hand-pumpers – but their rivalry was wholly friendly. Rivalry there was, however, with the regular firemen, who regarded the AFS with cold disdain, thoroughly suspicious of the motley collection of its members until those members had proved themselves. But the members of the AFS found it hard to

sustain their enthusiasm for long. As London waited through the cold winter of 1939 for the first air raid, and waited on in vain until the spring, the fervour of the AFS and the tolerance of the public, forced to maintain them, were both sadly stretched. It was hardly surprising, therefore, that by the summer of 1940, the strength of the AFS had shrunk to about half its original numbers, as the men left to do other war work or to join the regular services.

When, finally, the shock of the Luftwaffe's attacks hit London in September, the majority of the AFS had never come into contact with a fire. They had to take the full brunt of the work and the danger. As the bombs poured down and fires broke out all over the city and in the docks, other people might try to take shelter, but the firemen could not protect themselves at all as they struggled to douse the flames that would act as beacons to the next wave of bombers. To add to their problems, water mains burst and water jets failed, just when they needed maximum output with which to fight the fires: it was estimated that as many as 100 million gallons of water might be used during one air raid. When the raids were at their worst, the effect they had was devastating and demoralizing: as many as 2000 fires might be started in a single night and the death toll often rose high into the hundreds. To tackle this sort of work, the AFS needed to prove themselves quickly, and they did so with courage and dedication that won them the gratitude of all Londoners.

Left: This ATV with its trailer pump was typical of wartime improvisation.
Bottom left: Even the London taxi was pressed into service with the Auxiliary Fire Service.
Below: 1956 Dennis F8, a traditional and highly successful British style.

London was not the only city to burn. Coventry, Liverpool, Plymouth, these and many others suffered terrible fires, which the firemen rarely had enough equipment to tackle. Nor was Britain the only country to suffer severe fire damage. When the tables were turned, even the efficiency and the precautions of the highly organized German fire departments could do nothing against the concentrated bomber attacks of the RAF on Hamburg, Essen and other cities, where fire storms engulfed everything and sucked into their destructive heat trees, houses and people. There was little the fire service could do but briefly play jets of water along narrow paths to allow momentary escape routes that were all too soon engulfed in the flames once more. No equipment, however versatile or extensive, could hope to contend with such a maelstrom of fire.

European machines of these years and the years immediately before the war included the chubby Dennis Ace of 1938, built for Sonning in Berkshire, and equipped with 1700 feet of hose, a 35-foot telescopic ladder, foam-generating apparatus and suction hose. Glasgow fitted out a pump appliance on an Albion CX14 chassis in the same year. Two years earlier, Crossley supplied a foam and CO_2 tender, provided with a 200-gallon water tank and four 80lb CO_2 cylinders, all on a 6×4 chassis. Crossley supplied many fire-fighting vehicles during the war and immediately prior to the war. Another, specifically designed for use by the air force, was the Crossley FE1, also on a 6×4 chassis, with a 300-gallon water tank, a 28-gallon liquid foam tank and four 60lb CO_2 cylinders.

On the other side of the battle front, there was, for example, the all-wheel-drive Opel Blitz, with a 75bhp engine, a 350gpm front-mounted pump, and a 176gpm portable pump. This vehicle had a fully enclosed body. Similar chassis design, with the easily recognizable grille and cab, characterized the wartime Opel Blitz PWT, which had an additional 440-gallon tank so that it could double as a street-watering vehicle.

Many of the wartime designs lingered on for some years after the war as nations, economically exhausted by the conflict, were unable to afford new designs and were prepared to make do with the improvisations that had proved themselves during such difficult times. But slowly the lessons learned in war had their effect, and fire-fighting forces began to modify and expand their equipment. Among the more conventional pump escapes was the Dennis F range, which was put into production soon after the war. This began with the Dennis F1, which was powered by a four-cylinder 70bhp petrol engine and was equipped with a rear-mounted, multi-stage turbine pump which had an output of 400gpm at 120lbs psi. This was a narrow version of the F range which was specifically designed for country lanes and villages with their twisting corners and encroaching hedges and walls: the

Above: This 1945 Fordson FoT appliance was in service at Rochester Airport at the end of the war.
Right: One of the latest appliances in the long and respected Dennis range, the 1974 Dennis F.109 PE, in service with the London Fire Brigade.
Below: 1964 Thorneycroft/Pyrene Mk7A FoT, in service with the Royal Air Force.

chassis was only seven feet wide. It was improved on by the F2, which was powered by a Rolls-Royce B80 eight-cylinder petrol engine. The Rolls-Royce B80 Mk X petrol engine was still being used in the F7, introduced in 1949. The F7 had an impressive acceleration of 0–60mph in 45 seconds and was a well-known and much-admired fire engine. More popular still was the F12, which had a similar engine to the F7 and was equipped with a 900gpm mid-mounted No 3 Dennis pump. A later variation came with a rear-mounted 1000gpm pump and with no escape. The F14 was designed to take the Metz turntable ladder of 100 feet.

The F range of Dennis appliances is still in production today and is in use by fire brigades all over the world. The series provides a choice between a Rolls-Royce B81 S.V. eight-cylinder petrol engine which produces 235bhp at 4000rpm, and a Perkins V8 diesel engine which produces 180bhp at 2600rpm. The range of chassis includes a water tender, a water-tender ladder, a pump-escape, an emergency tender and a 50-foot Simon Snorkel. Special Purpose Appliance chassis are available for snorkels of greater length or for turntable ladders up to 125 feet.

The smaller Dennis D series retains the seven-foot width of the early F1 and remains ideal for small towns and country villages. Engine options are the Rolls-Royce B61 petrol engine and the Perkins

T6.354 diesel engine. The pump usually provided is the Dennis No 2 gun-metal pump with capacity output of 500gpm. Water tanks come with capacities of 100–400 gallons. Two first-aid hose reels up to 240 feet per reel are fitted each side, low behind the rear wheels, in a similar way to the F series. The same variety of bodywork is available as for the F series, with the exception of the snorkel chassis.

These may be said to be a fairly typical range of conventional pumpers. Dennis also provided airfield crash tenders, which became increasingly important after the war as aircraft grew rapidly in size with a consequent increase in the dangers from fuel explosions. It was, however, in America, immediately after the war, that the most important new developments occurred in the form of foam monitors for spraying onto airfield fires, using a

large tank with up to three tons of CO_2 and a special foam compound. This was sprayed onto the fire through several nozzles, including an overhead boom and a ground-level spray. The equipment was developed by the American Cardox Corporation. Examples were mounted on 6 × 6 Sterling DDs 235 or Reo 29–FF chassis.

Another important name in association with airfield tenders has been that of Pyrene. Dodge and Bedford chassis were used after the war for Pyrene equipment. One of the earliest Pyrene tenders was based on a Bedford QL Model all-wheel-drive chassis, in 1947. This carried a 500-gallon water tank, 30 gallons of foam compound and CO_2 extinguishing apparatus, with a 75hp Coventry Climax pump capable of an output of 400gpm. As much as 1800gpm of foam could be produced by the larger

Left: The Reynolds Boughton-Pyrene Pathfinder in action. This foam crash tender was first introduced in 1971 and has seen wide service.
Bottom left: A close-up view of the Pathfinder.

Right: The snorkel offers a commanding position from which to control fire and over reach obstacles.
Below: Snorkel Squad No 1 of the Chicago Fire Department – a Hendrickson/Pierce apparatus with a 55-foot boom.

126

Left: 1973 Oshkosh crash wagon, assigned to Chicago O'Hare International Airport. This is a 1000gpm pumper, carrying 1000 gallons of water and 1000lbs of Purple 'K',
Centre left: Yankee/Walter crash/foam truck in service at Los Angeles.
Bottom left: Crash/water truck with a Caterpillar tractor.
Right: An ex-Air Force crash truck at San Jose Airport.
Below: 1974 Yankee/Walter crash wagon, in service in Chicago. The examples on this page illustrate a new breed of fire-fighting appliance made necessary by the increasing size of airport fires.

Scammell-Pyrene foam tender, which could fight fires effectively from a safer distance. Whereas the fixed foam tender could only be stationed at one point in order to fight a fire, trailer-mounted monitors could be manoeuvred around the scene of the fire quickly to provide attack from different points, often simultaneously. Pyrene also built airfield crash tenders for the versatile Thorneycroft Nubian 4 × 4 chassis, which had seen very successful service during the Second World War. Other manufacturers have used Thorneycroft chassis to equally good effect.

In the 1970s, airfield crash tenders have become even more important and, of necessity, have incorporated some of the most sophisticated forms of fire-fighting equipment in existence. The name of Pyrene is still eminent. For example, there was the Pyrene Pathfinder of 1971, with a Reynolds Boughton six-wheel-drive chassis powered by a General Motors V16 supercharged two-stroke diesel engine of more than 600bhp. This vehicle had a 3000-gallon water tank and a 360-gallon liquid foam tank, with a Coventry Climax centrifugal pump that could deliver 1500 gallons per minute. The roof monitor was remote-controlled and could produce a jet of expanded foam which could reach as far as 250 feet at the rate of 13,500gpm. There was also provision for an alternative 1000gpm expanded-foam output through each of four sidelines. The ruggedly built vehicle could reach a speed of 60mph and provided the driver with complete visibility by setting his seat in the centre of the vehicle, rather than to one side, as is normal; fellow crewmen sat behind the driver.

Still more impressive is the Faun 8 × 8 airfield crash tender, which has a 1000hp engine and was built specifically with the prospect of coping with a Jumbo Jet crash containing 490 passengers and vast fuel tanks in the wings. For speed and control, the Faun has an eight-wheel all-drive rigid chassis. Of the two variations available, the foam tender can carry nearly 4000 gallons of water and 440 gallons of foam, and the powder tender can carry more than 2500lbs of dry powder.

As with the Pyrene Pathfinder, General Motors supplied the engine for the Kronenburg 6 × 6 airfield tender of 1972. This USA/Dutch cross had a six-wheel-drive chassis, a V8 two-stroke diesel engine for propulsion and a GM V6 petrol engine for the 440gpm pump. It was provided with approximately 1300 gallons of water and 250 gallons of foam.

House fires, factory fires, forest fires, airfield fires – these are only some of types that have to be contended with: there are usually machines to deal with all eventualities, each with its adaptations

Right: Gatcombe fire tug of Esso Fawley, equipped with snorkel and platform.
Below: Gloster-Saro 'Javelin' foam tender designed in 1979, capable of discharging 10,000gpm.

Above: Simon Snorkel in service with the Kent Fire Brigade.
Right: On both sides of the Atlantic, the snorkel has many uses and comes in various forms. This American appliance is in service at North Lake, Wisconsin,

Below: 1972 Young Crusader chassis with 85-foot Pitman snorkel in service with King's Park Fire Department, New York.

for the problems of a particular type of fire. Some machines are more adaptable than most to a variety of work, and chief among these must be the elevating platform or 'snorkel', which has become one of the most important arms of any major fire-fighting unit. Although elevating platforms have not wholly replaced the aerial ladder and the old water tower, they are a versatile development from these appliances and may often be used instead of them. The main advantage of the elevating platform is that it can be used to fight a fire from above and to sweep over the face of a building with the greatest ease. It can also be used for rescue work, providing both a sense of security to those being rescued and making it a lot easier for the rescuers to lower injured people. Several people can be fitted into one basket and the basket can be raised and lowered in general more quickly than on a ladder. Elevating platforms were at first adapted from tree-trimming platforms, which needed a substantial range of movement.

In 1958. the Pitman Manufacturing Company of Missouri produced an elevating platform for the Chicago Fire Department. The platform was 50 feet high and immediate tests very quickly proved its worth and its ability: in terms of firepower it could produce a jet of 1200gpm through a two-inch diameter nozzle at 100lbs psi. Later that year, the snorkel again proved itself, this time in a real fire, the notorious Chicago School Fire of December 1958. Ninety-two children and three teachers died in the fire – many of the children were found dead at their desks – but the snorkel did much to avert an even greater catastrophe. Shortly afterwards, snorkels proved their adaptability by rescuing people from a train crash in Chicago. It soon became clear that the snorkel also lessened the dangers to the firemen: they could avoid collapsing walls and

Above: Aerial ladders, like the snorkels, come on many different chassis according to the special demands of individual fire departments. This is a Ward LaFrance Ambassador Aerial Ladder at Santa Rosa, California.
Top left: 1974 Maxim 100-foot Aerial Ladder, diesel engine, with Providence, Rhode Island.
Left: 1971 Mack Tower Ladder, 75-foot, in service at Oceanside, New York.

could use the platform as a base from which to fight a fire through the windows of a burning building.

The hydraulic platforms themselves are made up of two booms hinged together, with a platform hinged to the top boom. The bottom boom is fixed to a turntable appliance on the chassis of the vehicle approximately above the rear wheels. The platform is fixed so that it remains horizontal at whatever angle the two booms are set. When at rest, the platform folds down on top of the vehicle with the hinged joint over the cab and the platform extending over the rear of the appliance. On some snorkels there is an extension arm hinged to the end of the top boom – a short arm to which the platform is attached. Hydraulic rams operate the booms and hydraulic motors rotate the turntable. For maximum control, there are controls at the base of the lower boom as well as in the platform itself, so that the crew in the platform can control their own

movement. Jacks fitted to the frame of the vehicle provide additional stability: unlike the aerial ladder, the range of reach of the hydraulic platform is always within the range of its stability, so that there need not be a safety-margin scale as on the aerial ladders. Between the platform and the base there is a communication link; there are one or two delivery hoses to the platform; there may also be a breathing-apparatus line from the base to the platform. Provision is also made for ladders to be attached to the booms so that, in an emergency, access can be gained to the platform while it is still elevated, although, of course, one of the major advantages of the elevating platform is that it can raise one or more crewmen instantly to the required level. An additional safety factor for the men on the platform is the provision of sprays beneath the platform which provide a curtain of water to protect the crew from the heat of the fire.

Heights of snorkel range from 45 to about 100 feet, although snorkels capable of reaching greater heights are in manufacture. The Simon Snorkel of 1965 had three articulated booms, at the end of which there was a cage that would carry up to six adults: this type already had the water-curtain nozzle for protecting those in the cage. Water was channelled to the platform monitor through a three-and-a-half-inch pipe. Movement of the booms and control of the jets could be made both from the

Above: This unique 1973 Seagrave 100-foot aerial, of Stillwater Ladder Company 1, is based on a chassis with steerable rear wheels.
Right: 1976 100-foot aerial, typical of the late 1970s appliance, in service at Topeka, Kansas.

Bottom right: This 1975 Seagrave 100-foot aerial ladder is with the City of Newark, New Jersey.
Below: 1973 American LaFrance Quint 100-foot aerial with 1000 gpm pump.

ground and from the platform. Maximum working height was 85 feet; horizontal reach was 41 feet. Simon Snorkels are made by Simon Engineering Dudley Company in England. Their first platforms were built in 1954, although specialization of the platform for fire-fighting purposes did not occur until 1961. There are connections on all models for hoses, breathing apparatus, power tools and other equipment, so that entry can be made at a high window from a platform with a hose for close work within the area of the fire. Current Simon models include the SS/220, with a maximum working height of 77 feet; the SS/263, with a working height of 91 feet; and the SS/300, with a working height of 103 feet. The three-boom system on which these models are based enables them to go up, over and down the other side of any obstruction, to land, for example, on the far side of a roof or within the confines of a balcony. Another advantage of these snorkels is that they can be mounted on varying kinds of chassis, for convenience and the requirements of the customer: for example, apart from the ordinary commercial chassis, they have been fitted to the Dodge K1050 and the Magirus 232 D19FI.

Other detailed aspects of the snorkel system include automatic stops at the limits of all cage movements; all piping totally enclosed for freedom

from accidental damage; a lifting eye provided underneath the cage to be used in removing wreckage; spotlights provided on the cage; provision for foam equipment to be operated from the cage, including the latest high-expansion foam generators with considerable output. Most of these snorkels can be at full working height and in operation within one minute of their arrival at the scene of the fire. For harbour work, there is in addition a versatile tug snorkel, which is fitted to the top of the tug and is in use in many parts of the world. Sweden is one of the major manufacturers of snorkel equipment. The 1966 Scania-Vabis Hydraulic Platform, based on the Scania LB 7650 chassis fitted with the Scania D 11 six-cylinder diesel engine, has a maximum horizontal reach of 45 feet and a vertical reach of between 70 and 80 feet. Meanwhile, in America, where the snorkel originated – it was nicknamed the 'snorkel' when it was seen appearing like a submarine snorkel above the smoke of a fire – the arguments as to its merits continued long and fiercely, as had the arguments about all previous innovations! But in time such famous names as American LaFrance and Mack produced their own versions of the aerial platform. The Mack variant came out under the name of the Mack Truck Aerialscope. This has a four-section extending boom fitted

Left: Mack tractor unit hauling New York Fire Department's Super Pumper with Napier Deltic engine and DeLaval centrifugal pump, 1965.
Below: 1969 Ward LaFrance Command Tower Pumper, with telescoping tower and 1250gpm pump.
Bottom: The Command Tower retracted for normal use.

INTERNATIONAL/ NATIONAL FOAM FIRE ENGINE, 1971

This unit is equipped with Ansul Dry Chemical and 200 gallons of light water. The apparatus is part of the fire fleet at the Mobil Oil refinery near Joliet, Illinois, which is one of the biggest installations of its kind. The increasing size and associated risks of such installations have necessitated the design of very sophisticated and specialist equipment. Parallel risks at international airports, where a very great percentage of a large airliner's payload consists of fuel, have created similar demand. In the same way, the fire risks of high-rise office blocks and crowded city streets have called for equally sophisticated equipment specializing in a different way. More and more, the story of fire engines is that of apparatus diversifying along specialist lines. The giant cannon on this foam pump can be seen pointing ahead over the cab of the truck.

with a fixed ladder, which is generally used only in an emergency: like the snorkel, it has a platform. On the Mack Big Reach 75-foot aerialscope, four booms of diminishing size fit one inside each other for compact storage on the truck, with the platform fixed to the smallest boom: the whole is fitted to a turntable and all pipes to the platform are enclosed within the booms themselves for protection from fire. This telescopic system is preferred over the articulated boom of the snorkel for reasons of manoeuvrability in narrow streets, particularly, for example, in New York City's financial district. Without moving, the Mack Big Reach can cover an area of 6,850 square feet up to a height of 65 feet and the boom can be elevated to an angle of 75 degrees from the horizontal.

Diagrams of the arcs of reach of snorkel platforms display elegant curves reminiscent of the neck of a swan or flamingo, but these machines, although they may appeal to the aesthetic eye, are wholly practical and have proved their usefulness again and again, on their own and while working alongside the remaining aerial ladders and water towers.

Right: 1973 Ford tractor with 5300-gallon tank and 300gpm pump, with Delaware City Fire Company.
Bottom right: 1977 Pierce 'Mini-Pumper' with a 250gpm pump – a increasingly popular appliance.
Below: Two kinds of firepower at the Watts Riots.

These are still very much in use – for example, Japan has its 1971 Hino six-wheel turntable aerial ladder, which has an extension of approximately 107 feet; the vehicle itself, with a six-cylinder diesel engine of 200hp, has a road speed of approximately 60mph. The ladder is in five sections. Hydraulic power for the lift is taken from the vehicle engine. There is a mid-mounted, two-stage turbine pump with an output of 500gpm at 120lbs psi and there are two outlets, on either side of the vehicle, and another mounted in a monitor at the head of the ladder. As in the hydraulic snorkel, there is a protective spray at the ladder head and also a searchlight.

The American Maxim FF-CLT 100-foot aerial ladder, also of 1971 manufacture, similarly shows that the aerial ladder still has its uses. This ladder is in four sections and is front-mounted over the vehicle chassis, although if required the appliance can also be manufacturered with a rear-mounted ladder. Comparable to this is the 100-foot Soviet AL-30 ladder truck: the ladder is in four sections and manual controls are provided for the three-man crew in the event of power failure.

The water tower finds its modern equivalent in the Ward LaFrance Command Tower of 1972. This can be fitted to a variety of normal pumpers to provide an elevated platform from which to fight smaller fires and is in part intended for those brigades that may not be able to lay out for the more expensive aerial ladders and snorkels. The Ward LaFrance tower can be raised hydraulically, with a two-man crew, to a height of 22 feet.

The output of all these appliances is dwarfed by what, at first sight, appears to be a pumper of a more conventional nature but is in fact the most powerful fire-fighting complex in the world: this is the dramatically titled Super Pumper Complex, developed for New York by Mack Trucks and put into service in 1965. In effect, it is nothing less than a land-based fireboat, capable of an endless gush of water. It has been realized for some time that what New York needed was something with the power of a fireboat that could be used on land, for many of New York's fires were too far from adequate water resources to receive the pressure that the pumpers really needed. If, for example, many pumps were drawing from the same water main, pressure inevitably dropped; reduced pressure meant not only that the jets of water did not reach as far as was required but that also the subsequently thinner stream of water was liable to vaporize completely even before hitting the fire. In addition, most pumps were too heavy and the hose was not strong enough to take the necessary pumping pressures that the firemen were seeking. It was naval research that led to

Right: Combating the effect of riots after the death of Martin Luther King.
Below: Chicago Fire Department's John Plant designed the powerful 3000gpm pumper, 'Big John'.

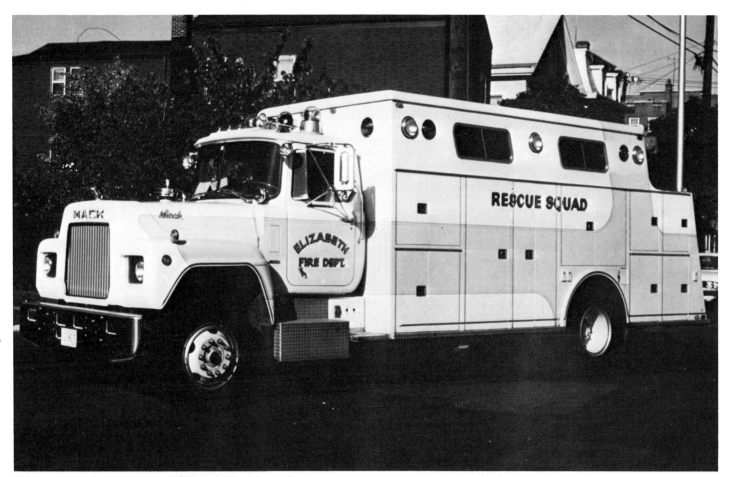

Above: Mack 'K' Rescue Squad Truck with Hannerly body, in service with the Elizabeth Fire Department, New Jersey.
Top left: One of the latest rescue trucks, this is a 1975 Ford model with the Odessa Fire Company.
Bottom left: Light rescue trucks are also useful for ancillary equipment.

some dramatic developments: the navy created lightweight diesel engines and lightweight, sturdy hose.

The Super Complex consists of six units: the Super Pumper itself; the Super Tender, with its vast water cannon; three Satellite Tenders, with smaller cannon; and a station wagon for the control officer. The Super Pumper has a Napier-Deltic 18-cylinder engine of 2400hp at 1800rpm, with an output potential of 8800gpm at 350lbs psi; for higher penetration, the pump can work at 4400gpm at 700lbs psi. The pump is a six-stage De Laval centrifugal pump which is made of stainless steel so that it can draw on either fresh or salt water. Total weight of the Mack F715ST tractor and trailer is 34 tons, loaded on to 18 wheels.

The Super Pumper is basically used in an intermediary stage as a mains pumper from the largest available supply of water, to pump water on to the tenders at the scene of the fire. In this capacity, drawing water from the waterfront or from a fireboat, the pumper can supply 37 tons of water a minute through 35 hose lines or through a lesser number of high-volume nozzles or through four master streams. The largest nozzle in the complex is the Big Bertha Stang Intelligiant cannon mounted on the Super Tender, which can send a jet of water 600 feet into a fire. The tractor unit that draws the Super Tender is similar to that which draws the Super Pumper. The water cannon has a capacity of 10,000gpm and is mounted just to the rear of the truck cab. Both the Super Pumper and the Super Tender carry 2000 feet of four-and-a-half-inch hose: standard hose has an internal diameter of two and a half inches only.

Each of the Satellite Tenders has a single water cannon amidships with a capacity of 4000gpm and each has 2000 feet of four-and-a-half-inch hose, similar to that of the two master vehicles. In action, one or more of the Satellites will usually arrive at the fire first, take up position, and will then be linked up to the Super Tender and the Super Pumper when they arrive. The Super Pumper may often be several blocks away from the scene of the fire at a predetermined water source.

Other cities have not been able to afford the outlay on such a system, but they have, instead, their own variations. Chicago, ever at the forefront of the fire-fighting battle, produced Big Mo in 1968, with two powerful nozzles, capable of directing a jet to the seventeenth floor of a high-rise block with a capacity of 1700gpm. Big John appeared a few years later, with hydraulically operated nozzles capable

MACK AERIALSCOPE, 1970s

This engine has a tandem rear axle, 325hp V-8 diesel engine and Mack 5-speed transmission. The rear axle is equipped with interwheel and interaxle power divider. When the centre of the turntable base is 32 feet from the building wall, the platform at the end of the telescopic boom has a contact area of 6850 square feet – far greater than that of conventional hinged boom apparatus. This makes the Aerialscope one of the most versatile pieces of equipment available to firemen. It can serve as water tower, observation platform and lift; the telescopic ladder that runs along the top of the boom can facilitate a mass rescue operation or a change of personnel in the platform itself without lowering the boom. Hydraulic stabilizers are situated in pairs at the front, rear and sides. They are shown retracted here. The pairs at front and rear lower directly to the ground, the centre pair stretching out to give the widest possible base and maximum stability.

of being elevated to 21 feet and able to throw a stream of water very nearly twice as far as New York's Super Tender.

Such are the giants of today. Sadly, the fires they have to fight are not all accidental. Many of the worst fires that must be tackled are the result of riots and arson; in such cases, fire fighters have to contend not only with the fire but also with the open antagonism of the rioters themselves, who are frequently anxious to prevent the firemen from fulfilling their duty. A typical challenge of this kind was faced by the District of Columbia Fire Department in the 1968 riots that developed after the death of Martin Luther King on Thursday 4 April.

The shock of King's death almost inevitably led the Negro community to use arson as a weapon of protest and as an escape valve for their feelings of frustration at the senselessness of the event. Arson began with minor incidents: a number of vehicles were overturned and set on fire. During the night, fires were started in several places, but the fire department had anticipated trouble of this kind and had initiated a set plan to alert as many firemen as possible, increasing manpower by calling men back from duty leave. However, the number of fires rapidly made it impossible for the department to extinguish them completely before moving on to the next blaze: instead, the fires were 'knocked down' so that they were no longer dangerous, thereby freeing the appliance to go to 'knock down' the next

fire, with the intention of returning later to make a final extinguishment. Fires were being raised at the rate of about 30 an hour and were being fuelled by gasoline thrown on by the rioters, so first-aid techniques of fire fighting were wholly inadequate. To assist in the crisis, surrounding communities contributed as many as 60 pumping appliances, as well as about seven aerial ladders.

By the early hours of 6 April, Army and National Guards troops were on the streets. By then, too, there was a curfew and a ban on the sale of alcohol and flammables such as gasoline, unless they went directly into the tank of a vehicle. By daytime, there had been more than 300 initial fires as well as more than 200 rekindlings of fires that had been temporarily 'knocked down'. By Monday, there had been 500 original fires and 300 rekindlings. The arrival of the military helped to deter those who were harrassing the fire fighters, but, nonetheless, firemen suffered many injuries not only from the fire but from objects thrown at them. Luckily, there were few fatalities.

Three years earlier, in August 1965, Los Angeles had faced worse riots in Watts, an area towards the south of the city. From the small beginning of a scuffle over a routine arrest, rumour had spread to unrest and riot throughout Watts, which had a 95 percent black population. Population density was high, even though the area did not immediately look like a bad slum ghetto. There was unemploy-

Above: A turntable ladder with folding rescue cage mounted on a Renault GF 231 chassis.
Left: Dennis 'RS' fire appliance of 1978 equipped with a 500gpm main pump.

ment, there were fights with police, there was discontent and disillusionment. In minor fires, objects had already been thrown at firemen, who had been promised protection by the police in attending further anticipated fires.

As in 1968, the tactics were to 'knock down' fires with task forces strong enough to deter attack against the fire fighters themselves, but the police escorts were not always where they were most needed. Bricks and rocks were thrown and apparatus was unable to move fast enough through the streets, impeded by looters and arsonists shouting, 'Burn, baby, burn!' Even the police chief admitted that the situation was out of control. Cocktail bombs were used to fire the buildings and looters held back the firemen until property had been thoroughly ransacked. At times, guns were used against the firemen.

With 200 fires burning at one point, there was a backlog of as many as 20 fires waiting for appliances, but alarms from fireboxes were often ignored because so many of them were false and were pulled especially to lure fire fighters into ambushes. Officers rode shotgun on some appliances, which were often preceded by National Guardsmen in jeeps with machine guns. In some cases, firemen were forced to abandon their equipment, which itself was then looted by rioters. The rioting continued for nearly three days, during which 261 buildings were destroyed, 34 people were killed, 4000 people were arrested and 180 fire fighters were injured. Two years later, in riots in Detroit, 1300 buildings were destroyed!

The pattern was constantly repeated, as fire

fighters found themselves as much the object of the rioters' hatred as the police or the government. Looters and rioters would repeatedly obstruct them in the course of their duty. As they fought to save one building, arsonists would run out and fire the next building. The tragedy has been that, too often, firemen have been regarded as symbols of authority when, in fact, their loyalties are impartial, their only duty being to save life and property.

This has always been their duty, throughout the changes in the equipment with which they have been supplied. The modern fire brigade is very different from that of a hundred or more years ago, for the small choice of hand pump and siphon has given way to a vast range of new equipment. But much of the equipment has one thing in common: it is hand-made to specific orders. This explains the great variety. If the demand for equipment were far greater, the supply might be produced on a much larger scale. But every fire department has specific needs which demand specific adaptations – and all fire equipment must be to the highest standards.

This is the moment to take a last look at the range of appliances available to the fireman of today. The main attack power is in the form of the pumper. Twenty years ago, pumpers of 750gpm capacity were the common order of the day; now there are 1000gpm and even 2000gpm pumpers. Of course, these are not always pumping at full capacity; pressure and water output are balanced against each other to achieve precisely the flow that is required for the particular occasion. A standard 1000gpm

Right: An American LaFrance pumper fire appliance. Below: A Ward LaFrance Ambassador with a telescopic water tower built in New York in the early 1970s.

pump producing a jet of 150lbs psi might produce 700gpm at 200 psi or 500gpm at 250lbs psi for a more penetrating jet.

Water tanks and water tankers, or 'combination' vehicles supplied with water, are common in areas where water supplies are not always readily available. The tank carried by these vehicles may have a capacity of anything between 300 and 1000 gallons. Separate tankers, with a minimal pumping capacity – enough merely to pump their water into the tank of the main appliance – might hold anything from 1000 to 1500 gallons of water.

The amount of hose carried will vary considerably; 1500 feet of two-and-a-half-inch hose might be considered average for a pumper. Hose is normally rubber-lined, with a double jacket of woven cotton; more recently, this jacket consists of polyester fibre or a mixture of cotton and polyester fibre. Polyester fibre has the advantage of being lighter and tends to rot less easily, but it does have the disadvantage that it may kink in use. One-and-a-half-inch and three-inch hose are also used, as well as up to four-and-a-half-inch hose on the Super Pumper Complex. Nozzle types vary between straight stream and spray (or fog) types.

American ladder trucks are still considered of the utmost importance, and truck men are often required to serve on a pump before they move to a truck so that they may understand just how vital a role the truck does play in association with the pump. The primary task of the truck men is to save life. The typical modern truck has a 100-foot hydraulic-lift all-metal aerial ladder; it is also equipped with up to 200 feet of ground ladders, its own water tank, a booster pump, hose, a generator and floodlights. The necessity to make forcible entry into buildings requires, in addition, such items as pike poles, pry bars, axes and power saws to hack through windows, roofs, ceilings and floors. Complementary to these major pieces of equipment are the snorkels, which we have already looked at in some detail.

There are also several supplementary appliances. For example, America has its squad-rescue units, which perform duties similar to those of the ladder truck. The squad-rescue unit brings the extra equipment needed by the truck company: it brings more power tools, breathing apparatus, power winches, resuscitators and other life-saving equipment. Some squad-rescue units have their own turret nozzle and pump.

The duty of the salvage-unit men is to rescue what stock they can, as much from water damage as from damage caused by the fire. In large fires, their work can save millions of pounds worth of property from being lost and their work is no less dangerous than that of other firemen. They are the sophisticated equivalents of some of those Roman fellows, 2000

years before, armed with brooms, shovels and mops, but now carrying modern water vacuums and salvage covers as well.

Dry chemical units are usually used in conjunction with foam units, since the dry chemical, although effective in blanketing a fire, does not always have the necessary cooling effects.

Crash wagons are common at airports. They must be fast and able to cross rugged terrain to reach off-runway crash sites. In general, foam-generating crash wagons at large airports carry their own water supplies of between 2500 and 3000 gallons, as well as 500 gallons of foam in two separate containers. Either hand-held nozzles can be used or nozzles mounted on separate vehicles; a main turret nozzle can also be used from the appliance itself.

Amphibious units are increasingly popular where large areas of water adjoin fire-risk areas. One such unit is the Alvis Salamander airfield crash tender, powered by a Rolls-Royce B81 six-cylinder petrol engine producing 240bhp; another is the Eisenwerke amphibious appliance in use with the Federal Republic of Germany, with a 12-cylinder diesel engine and a road speed of approximately 56mph. This is capable of crossing grassland, marshland, open river and of climbing steep banks. It has two 880gpm centrifugal pumps.

Smoke ejectors, fireboats, electrical generating units, command units, half-track and full-track

Above: Amphibious fire-fighting appliances often find short cuts to otherwise inaccessible fires. Left: Designed in 1976, the HCB-Angus CSV type 'B' water tender set new standards of crew safety.

forest-fire vehicles, jeeps, helicopters, ambulances, canteens – these are all part of the armoury of the fire fighter, the equipment with which he must tackle his adversary. They all have their place in his strategy and tactics, however much it is the front-line pumpers that first catch our interest in this fascinating display of specialized machinery.

There is no doubt that the engines of the future will be developed out of all recognition, even though the main emphasis today is on preventing the fire before it happens. Computerization will slowly take hold of the fire fighter's life, as he battles to save the thousands of people that cram into structures 100 storeys high, or fights as hard as he has always had to do to save a family from their burning home – the most common source of fires. No doubt we will be as surprised, in 50 or 100 years' time, as our ancestors would have been to see us 50 or 100 years on. In reflecting on the future of the fire engine, we should remember that only 300 years ago fire-fighting appliances were on the whole as unsophisticated as, and sometimes a great deal less sophisticated than, they had been in Roman times. What might happen in the next 300 years is anyone's guess!

Glossary

Aerials Fire ladders generally based on turntables, which originated in the late nineteenth century with those produced by Hayes, Dederick and Seagrave.

Air vessel Airtight chamber that contains air at atmospheric pressure. When fitted to the outlet of a water pump, it transforms a pulsating jet of water into a continuous stream.

Appliance General term used for any type of fire-fighting vehicle.

Aquarius Man in charge of fetching the water in the Roman Corps of *Vigiles*.

Braidwood body Design of fire engine on which the crew facing outwards sat on either side.

Bucket and plunger A combination of both 'force' and 'lift' pumps, similar to the double-acting force pump.

Centrifugal pump Revolving pump in which water drawn in at the centre is expelled at the periphery by centrifugal force.

Chemical engine Chemical apparatus or rig in which carbon dioxide is used to expel water from a container under pressure.

Cider Mill Manual pump using a windlass to provide a rotary action within the pump and drive a wheel fitted with teeth that scoop the water down a pipe.

Coffee mill Pump similar in action to the cider-mill pump but with a handle at the side instead of a windlass on top.

Combination appliance A fire engine which combines at least two functions, such as those of main pump and ancillary tender, or main pump and ladder escape. Sometimes called a dual-purpose appliance.

Corps of Vigiles Highly organized, paid Roman fire-fighters, instituted by the Emperor Augustus. There was one cohort of 1,000 men for each of the seven areas of Rome.

Curfew From the French *couvre feu* or 'fire cover'. It referred to the time at which all fires in a town were to be covered for safety during the night.

Double-acting pump Type of force pump or reciprocating pump with a single cylinder in which the piston forces water out on both the up and down stroke.

Double cylinder pump Pump in which a piston stroke is made in each of the two cylinders alternately; one piston draws water into its cylinder as the other thrusts it out.

Double deckers Manual pumps on which there were two rows of men on either side. One row stood on the ground, the other on the appliance itself.

End-pumper Manual pump with the handles fixed at either end, instead of at the sides.

Escape Ladder used to reach upper windows and roofs of buildings. Ladder escapes could be free standing, fitted to motorized appliances or fitted in combination with pumps.

Familia Publica Roman fire-fighters, usually slaves who preceded the Corps of *Vigiles*.

Fireboat Vessel fitted with pump or pumps and normally used at docksides.

First-aid vehicle Vehicle fitted with emergency equipment such as hand extinguishers, chemical extinguishers, short ladders, axes and so on. It is equipped to reach a fire as quickly as possible and begin fire fighting before the main pump arrives.

Fly ladder An extension ladder attached to the top of the main ladder and swung up by ropes.

Foam tender Fire-fighting appliance especially adapted for use at such places as airports where oil fires require blanketing with foam.

Gooseneck Swivel-jointed tube in discharge pipe of manual appliance, enabling a jet of water to be directed accurately.

Gpm — Gallons per minute.

Haywagon — Nineteenth-century manual pump with double sets of handles that folded up to give it the appearance of a haywagon.

Hook and ladder — American appliance, equipped basically with long ladders and hooks, to act in conjunction with a pump appliance.

Hose wagon tender — Appliance for carrying lengths of hose.

Manuals — Pumps operated by hand.

Mobile dam — Trucks converted for use as water carriers, especially during the Second World War.

New World body — Design of fire engine on which the crew sat on either side facing inwards.

Piano-style — Pumps so-called because of the shape of their water boxes.

Pump escapes — Popular in Britain, these are a combination of a pump appliance with an escape ladder, characterized by the ladder's large wheels at the rear.

Preventers — Early fire hooks used for dragging burning material from roofs or for pulling down houses in order to create a path for the fire and then prevent it from spreading in all directions.

Psi — Pounds per square inch, pressure.

Rattle watch — Volunteer fire groups in the early years of North American settlement. They kept watch in the towns at night and warned local people of fire by sounding their rattles.

Reciprocating pump — The basic style of pump as used in manual appliances, working in a to-and-fro motion.

Rotary pump — Pump providing a continuous flow of water, without an air vessel, by means of inter-locking, toothed wheels revolving in opposite directions. Water is drawn in as they revolve and forced around their perimeter and from there out of the pump.

Rpm — Revolutions per minute of the pump engine.

Salvage unit — Firemen equipped to rescue stock, both from fire and the water of the appliances.

Sapeurs Pompiers — French fire-fighting company instituted by the Emperor Napoleon.

Shanghai — Nineteenth-century manual pump with pagoda-like decking.

Side-pumper — Manual pump with handles on either side, as opposed to the ends.

Single-acting pump — Simple pump in which the water is forced out by the piston either on the down stroke or on the up stroke but not on both.

Siphonarius — Man in charge of the pump in the Roman Corps of *Vigiles*.

Siphos — Old name for water pumps, used by Hero of Alexandria.

Snorkel — Hydraulically elevated platform on two or three hinged booms, fixed to a turntable appliance.

Squirrel tail — Nineteenth-century manual in which the suction hose, when not in use, curved back over the rig like a squirrel's tail.

Squirt — Old name for simple water pump, in which a cylinder for holding water was fitted with a nozzle at one end and a plunger at the other. One man held a handle on either side of the cylinder while a third pressed down the plunger.

Syringe — Another name for a squirt.

Tender — Appliance to back up the main pump. A tender could be for water, men, equipment and so on.

Tractor — Name sometimes used for motorized unit.

Turntable ladders — Ladders fixed to a turntable base. These are the European equivalent of the American aerial ladders.

Uncinarius — The man in charge of the hooks or preventers in the Roman Corps of *Vigiles*.

Water tower — Scaffolding-like tower fitted to moveable appliance from the top of which water could be directed onto the fire.

Bibliography

BLACKSTONE, G. V. B., *A History of the British Fire Service*, Routledge and Kegan Paul, London 1957.

BRAIDWOOD, JAMES, *On the Construction of Fire Engines and Apparatus*, Edinburgh, 1830.

COLBURN, ROBERT. E., *Fire Protection and Suppression*, McGraw Hill, Berks & New York 1975.

CREIGHTON, JOHN, *Fire Engines of Europe*, Ian Henry Publications, Hornchurch, 1980.

DA COSTA, P., *100 Years of America's Fire-Fighting Apparatus*, New York, 1964.

DITZEL, PAUL, *Fire Engines, Firefighters: The Men, Equipment & Machines from Colonial Days to The Present*, Crown Publishers Inc., New York 1976.

EYRE AND HADFIELD, *The Fire Service To-Day*, OUP, London, 1944.

GAMBLE, SIDNEY, *Outbreaks of Fire*, C. Griffin & Co., London, 1931.

GILBERT, K. R., *Fire Fighting Appliances*, Science Museum: HMSO, 1969.

GLORY, C. O., *100 Years of Glory*, District of Columbia Fire Department, 1971.

HASS, ED., *Ahrens-Fox: 'The Rolls-Royce of Fire Engines'*, Ed Hass, Sunnyvale CA 94086, 1983.

HER MAJESTY'S STATIONERY OFFICE, *Manual of Firemanship*, HMSO, 1973.

INGRAM, ARTHUR & BISHOP, DENIS, *Fire Engines*, Blandford Press, Dorset, 1973; Macmillan, New York, 1977.

JACKSON, WILLIAM E., *London's Fire Brigades*, Longmans, Essex, 1966.

MALLET, J., *Fire Engines of the World*, Osprey, London, 1982.

McCALL, WALTER, *American Fire Engines Since 1900*, Crestline Publishing Co. Ltd., Florida, 1976.

OLYSLAGER ORGANIZATION, *Fire and Crash Vehicles from 1950*, Frederick Warne, London, 1976.

OLYSLAGER ORGANIZATION, *Fire Fighting Vehicles 1840-1950*, Frederick Warne, London, 1982.

ROETTER, CHARLES, *Fire is their Enemy*, Angus and Robertson, London, 1962.

YOUNG, CHARLES, *Fires, Fire Engines and Fire Brigades*, Lockwood, London, 1866.

Acknowledgements

Photographs
Endpapers Brown Brothers; Title page Zefa; Half title page Photri; Contents page Radio Times Hulton Picture Library; 6–7 Photri; 9 Bettmann Archive; 10–11, 11 Photri; 12 Bettmann Archive; 12–13 Brown Brothers; 17 Bettmann Archive; 18 Alex Langley/Aspect Picture Library 19 Photri; 20–21 Brown Brothers; 22–23 Photo. Science Museum, London; 24 Ann Ronan Picture Library; 25 Photo. Science Museum, London; 26 Bibliothèque Nationale/Snark International; 27 British Museum/Michael Holford Library; 28 top G.L.C. London Fire Brigade, Photographic Service; 28 bottom Radio Times Hulton Picture Library; 29, 30 Photo. Science Museum, London; 31 G.L.C. London Fire Brigade, Photographic Service; 32, 33 Museum of London; 34–35 Jack C. Novak/Photri; 37 Courtesy, I.N.A. Corp. Museum; 38 bottom Crown Copyright, Science Museum, London; 38 top Courtesy, I.N.A. Corporation Museum; 39 Edward R. Tufts; 40 Courtesy, I.N.A. Corporation Museum; 41 Library of Congress; 42 Museum of the City of New York/Scala; 42–43 Fire Museum of Maryland; 43 top Hall of Flame Collection; 44 Courtesy, I.N.A. Corporation Museum; 46–47 Crown Copyright, Science Museum, London; 48 Radio Times Hulton Picture Library; 49 Photri; 50–51 Fire Museum of Maryland; 52 Photo. Science Museum, London; 53 G.L.C. London Fire Brigade, Photographic Service; 54 Ann Ronan Picture Library; 55 top G.L.C. London Fire Brigade, Photographic Service; 55 bottom Crown Copyright, Science Museum, London; 56 Photo. Science Museum, London; 57 top G.L.C. London Fire Brigade, Photographic Service; 57 bottom Photo. Science Museum, London; 58 Brown Brothers; 59 top library of Congress; 59 bottom Courtesy, I.N.A. Corporation Museum; 60 bottom G.L.C. London Fire Brigade, Photographic Service; 60–61 Library of Congress; 61 Popperfoto; 62 Fire Museum of Maryland; 65 Photri; 66–67 Fire Museum of Maryland; 68 top California Historical Society Library; 68 bottom, 69 Courtesy, I.N.A. Corp. Museum; 72 Brown Brothers; 73 American La France; 74, 74–75 Fire Museum of Maryland; 75, 76 Radio Times Hulton Picture Library; 76 bottom G.L.C. London Fire Brigade, Photographic Service; 77 Courtesy, I.N.A. Corporation Museum; 78 Fire Museum of Maryland; 78–79, 79 Fire Museum of Maryland; 80 top Nevada Historical Society; 80 bottom Waterous Company; 81 Mary Whalen Collection in the Archives, University of Alaska, Fairbanks; 82–83 Mack Trucks Inc., 84 National Fire Museum, Newton Highlands; 85 Photo. Science Museum, London; 88 Waterous Company; 88–89 Radio Times Hulton Picture Library; 89 top right Protector Lamp and Lighting Co. Ltd.; 89 Waterous Company; 91 Tüzoltó Múzeum, Budapest; 90–91 Fire Museum of Maryland; 92 National Fire Museum, Newton Highlands; 92–93 Fire Museum of Maryland; 93 Melbourne Fire Brigade Historical Society/Colourviews; 94 Bob Graham; 94–95 Fire Museum of Maryland; 95 Neill Bruce; 96 Hestair Dennis Ltd.; 97 top Osterreichische Nationalbibliothek; 97 bottom G.L.C. London Fire Brigade, Photographic Service; 98–99 Roger Mardon; 99 Andrew Morland; 100 top Radio Times Hulton Picture Library; 100 centre Osterreichische Nationalbibliothek; 100 bottom Photri; 100–101 G.L.C. London Fire Brigade, Photographic Service; 101 top Photri; 101 bottom Radio Times Hulton Picture Library; 102, 103 Bob Graham; 104 Radio Times Hulton Picture Library; 105 Brown Brothers; 108 G.L.C. London Fire Brigade, Photographic Service; 109 Photri; 110, 110–111 Bob Graham; 112 Photo. Boyer/Roger Viollet; 113 top Photo. Science Museum, London; 113 bottom Photri; 114 Bob Graham; 115 Edward R. Tufts; 116 Photri; 116–117, 117 G.L.C. London Fire Brigade, Photographic Service; 118–119 Bob Graham; 120 G.L.C. London Fire Brigade, Photographic Service; 121,122, 123 Roger Mardon; 124 top Chubb Fire Security Ltd.; 124 bottom Bob Graham; 125 top Simon Engineering Dudley; 125 bottom, 126 top, 126–127 Jeff Schielke; 126 centre, 126 bottom, 127 top Bob Graham; 128 bottom Roger Pennington; 129, 130 Roger Mardon; 130–131 top Jeff Schielke; 130–131 bottom, 132 top Roger Mardon; 132 centre Bob Hatter; 133 Ward La France; 134 bottom Bob Graham; 135 bottom Roger Mardon; 134 top, 134–135 Bob Graham; 136–137 Mack Trucks Inc.; 137 centre Ward La France; 137 bottom Jeff Schielke; 140 Larry Schiller/Magnum from John Hillelson Agency; 140–141 Roger Mardon; 141, 142 Jeff Schielke; 143 Burt Glinn/Magnum from John Hillelson Agency; 144, 145 Roger Mardon; 148, 149 Roger Pennington; 150–151, 151 Andrew Morland; 152 top Roger Pennington; 153 Eisenwerke Kaiserslautern.

The Manufacturers

Agnew, John: of Philadelphia, built about 150 manual end-pumpers from 1823. These were among the most powerful hand pumps of the time.

Ahrens Manufacturing Company: purchased in 1868 by Ahrens, an employee of Lane and Bodley, which had bought the original Latta Company in 1862. Produced about 300 steam fire engines.

Ahrens-Fox Fire Engine Company: founded in 1908 in Cincinnati by Ahrens and Fox after both had withdrawn from the American LaFrance Company. Produced their first petrol-driven engine in 1911 and an aerial ladder in 1923. Engines characterized by the dominant spherical air vessel set over the pump at the front of the appliance.

American Fire Engine Company: formed in 1891 at Seneca Falls, Elmira, New York State from the amalgamation of Button, Silsby, Ahrens and Clapp and Thomas Manning Company in 1900 to form the International Fire Engine Company, which became the Amercian LaFrance Fire Engine Company in 1903. Seneca Falls was sometimes known as the 'Fire Engine Capital of the World'.

American LaFrance: formed in 1903 after various amalgamations. Major manufacturer of pumpers, aerial ladders and snorkel-type extensions.

Amoskeag: of New Hampshire. One of their first engines was shown in New York in 1860 and they subsequently became one of the major manufacturers of steamers.

Babcock Manufacturing Company: manufactured wheeled chemical engines and aerial ladders from the second half of the nineteenth century.

Bedford: chassis used as basis for armed forces fire-fighting appliances and in conjunction with Pyrene as foam crash tenders at airfields. Modern TK truck range introduced in 1960 and used widely.

Braithwaite, John: constructed the first steam fire engine in Europe in 1829, with the help of Ericsson. This was a two-cylinder, ten-horsepower appliance.

Braun: produced electric fire engines at the beginning of the twentieth century in Nuremberg.

Button, Lysander: built about 500 hand-pumpers at Waterford, New York, from 1834.

Christie, John: specialized in front-wheel-drive tractors in the American market at the beginning of the twentieth century; used for conversion of horse-drawn appliances.

Clapp and Jones: popular steamers in the second half of the nineteenth century.

Crossley: produced air-force tender and other fire-fighting appliances for the armed forces; in great demand during the Second World War.

Daimler: produced his first fire engine with an internal combustion engine in 1888. This engine still relied on horses to draw it.

Delahaye: automobile manufacturers with a history beginning at the end of the nineteenth century; subsequently produced large numbers of fire appliances for the *Sapeurs Pompiers* of Paris and the provinces.

Dennis: begun by two brothers who made bicycles in the 1890s and produced their first fire engine in 1908 which they sold to Bradford Fire Brigade. Became very popular with their Dennis Big 4 and Big 6 appliances and the modern F range and smaller D range of pumps. Bodies also used to carry snorkels and as crash tenders.

Faun: the Faun 8x8 airfield crash tender is especially designed for big Jumbo Jet crashes, although Faun have been involved in fire fighting since the nineteenth century.

Ford: introduced the Model T in 1908, which was soon popular as a fire chief's vehicle. Subsequently the chassis was widely used as a base for small fire engines. Later chassis also used for appliances.

Gorter, Henry: famous for water towers, built at the end of the nineteenth century. Among these was a 65-foot water tower built for San Francisco in 1898, which was subsequently motorized.

Hayes, Daniel: built the first successful aerial ladder truck; the ladder had to be cranked up by six men. The 85-foot wooden aerial ladder was built for San Francisco in 1868 and had a turntable.

Heyden, Van der: manufactured hand pumps in Holland and wrote the first handbook on fire engines in 1690. Introduced the hosepipe which revolutionized techniques by enabling fire fighters to get in close to a fire.

Hino: one of the several modern Japanese manufacturers. Among its products is the 32-metre turntable ladder of 1971.

Hodge, Paul: built the first self-propelled fire engine (and the first steam fire engine to be built in America) in 1840. It was called the Exterminator.

Hunneman, William: of Boston, one of the most popular hand-pump manufacturers. From 1792, the firm built more than 700 hand-pumpers. Subsequently they built a limited number of steamers but were outclassed by other manufacturers.

Kronenburg: involved in fire fighting since the nineteenth century; now produce a range of appliances from airfield crash tenders to foam and dry-powder appliances.

LaFrance: joined with American Fire Engine Company to become American LaFrance in 1903. Begun by Truckson Slocum LaFrance, who found work as an iron-founder in Elmira and subsequently became interested in fire engines and improved rotary engines.

Latta and Shawk: primarily under the direction of A. B. Latta, who built the Uncle Joe Ross for Cincinnati in 1852, in response to a prize offer for a successful steam fire engine.

Laurin and Klement: Czechoslovakian firm which began with bicycles in 1895 and moved through motor cycles to a variety of vehicles including fire engines. Merged with Skoda in 1925.

Lee and Larnard: manufactured in New York, these engines attended trials at Philadelphia, as well as trials at the International Exhibition in London in 1862, for steam fire engines.

Leyland: began as manufacturers of steam engines for trucks and lawn mowers. They turned to petrol motors and in 1910 delivered their first petrol fire engine to the

Dublin Fire Brigade at the special request of the Chief Officer. The hugely successful FE series of the 1920s was followed by the FK, TLM and FT ranges of the 1930s. Later combined with Metz to produce turntable ladders.

Lote, Thomas: manufactured what was possibly the first manual fire engine built in New York, in 1743. It was known as Old Brass Backs. Lote himself was a cooper and boat builder.

Lyon, Pat: manufactured about 150 hand-pumpers at the beginning of the nineteenth century for sale around New York, Philadelphia and Boston. These were mostly end-pumpers.

Mack: the five Mack brothers, from Brooklyn, started in the motor bus business in 1900, produced their first hook and ladder truck nine years later and their first pumper in 1911. Best known for the Mack Bulldog, or Model AC, of 1915. Characterized by letter M on the front of their tractors. Also responsible for the Mack Aerialscope and the Super Pumper of 1965 for New York.

Magirus: German firm which introduced the first motorized turntable ladder in 1892.

Magirus-Deutz: now one of the leading specialist manufacturers of airfield crash tenders, pumpers and aerial ladders.

Mason, Richard: began to build end-stroke pumpers in Philadelphia in the 1760s and was in business with his son, Philip.

Maxim: began in 1914 in New England with fire trucks, which were later followed by aerial ladders. Control taken over by the Seagrave Corporation in 1950s but a range of appliances continued to be manufactured under the Maxim name.

Merrick: American steam engine manufacturers whose Weccacoe took part in the 1858 Philadelphia trials.

Merryweather: one of the most respected British names in fire-engine manufacture. Moses Merryweather took over the firm of Hadley, Simpkin and Lott, which he had joined as a mechanic, after marrying the niece of Henry Lott in 1836. Hadley himself had first taken over a fire-engine works originally started by Adam Nuttall in about 1750. In 1861, Merryweather produced their first steamer, the Deluge, followed in 1863 by the Torrent and the Sutherland. In 1899, the firm produced its first self-propelled steamer, Fire King, and in 1904 it claimed the first self-propelled motor pump to be used by a public fire brigade.

Metz: the firm began in the 1840s and produced its first turntable ladder in 1812. Contracted with Leyland to produce Leyland-Metz turntable ladders. Continue to produce range of appliances from pumps to airport crash tenders.

Newsham, Richard: first patented an improved fire engine in 1721. His hand-pumpers re-introduced the air-vessel to promote a continuous stream of water and had a powerful influence on early eighteenth-century manufacturers. Handles were placed at the sides of the engine to make room for a greater number of men and further men were able to work the engine on treadles fitted in the centre of the engine.

Pirsch, Peter and Sons: founded in 1900 by Peter Pirsch, who had started off in the Nicholas Pirsch Wagon and Carriage Plant, Wisconsin. The 1895 Pirsch hook and ladder wagon was an example of their work. Starting with extension ladders, Peter Pirsch moved

through chemical engines to combination pumpers in 1916. Now manufactures pumpers, aerial ladders and snorkels.

Poole and Hunt: built their first steamer in 1858. Their Baltimore and Washington took part in the 1858 Philadelphia trials. Their output was relatively small: they built only seven engines in eight years.

Reaney and Neafie: manufacturers of steam engines; their Good Intent, Hibernia and Mechanic all took part in Philadelphia trials.

Roberts, William: built the first self-propelled steam fire engine in Europe in 1862; it was a three-wheeled monster weighing seven-and-a-half tons. His second steamer was the Princess of Wales.

Rodgers, John: Baltimore manufacturer of end-stroke pumpers during the first half of the nineteenth century.

Rosenbauer: Austrian firm with a wide variety of fire-fighting appliances; in existence since the second half of the nineteenth century.

Scania-Vabis: Swedish firm originally begun in 1891 as a railway repair shop, subsequently began producing motor vehicles as early as 1897 and later included engines in its range of products. Modern appliances include hydraulic platforms.

Seagrave: founded in Columbus, Ohio, in 1907. Started with chemical engines and moved on to pumpers and aerial ladders, for which the firm is most famous.

Shand Mason: the firm was created in 1851 but originated in 1774 with a mechanic called Phillips who built Newsham-type engines. In 1798 Hopwood succeeded Phillips; in 1820 the firm was taken over by Tilley and subsequently by Shand Mason, who built their first steam engine in 1858.

Silsby: another firm from Seneca Falls, Elmira, which started in 1845, and became famous for its steamers. Silsby made a rotary engine for New York City in 1861. Later they became part of the American Fire Engine Company and subsequently part of American LaFrance.

Simon: major British manufacturer of snorkel equipment, based on a variety of chassis.

Skoda: Czechoslovakian firm which benefited from a merger with Laurin and Klement early this century and now produces highly versatile pump appliances.

Smith, James: manufactured about 500 goosenecks in the New York area from 1812.

Teudloff-Dittrich: Hungarian manufacturers of manual fire engines in 1885; subsequently produced steamers and some interesting combinations of petrol-powered pumps with horsedrawn chassis.

Thayer, Ephraim: began manufacturing hand-pumpers in Boston in 1794.

Thorneycroft: major manufacturer of aircraft crash tenders; the Thorneycroft-Nubian 4 x 4 chassis has been superseded by the Nubian Major.

Ward LaFrance: originated in Elmira in 1918; renowned for ladder trucks; currently producing a range of equipment including command towers for fitting onto existing appliances.

Waterous Engine Works: from St Paul, Minnesota. Rivalled American LaFrance for the first motor-driven apparatus in America. Manufactured a horse-drawn, petrol-powered pumper in 1898. In 1907, manufactured a pumper with a single engine that provided power to the pumps and to the road.

Index